Loft Conversions

John Coutts

MA (Oxon)

Blackwell
Publishing

Blackwell Publishing Ltd
Editorial offices:
Blackwell Publishing Ltd, 9600 Garsington Road, Oxford OX4 2DQ, UK
 Tel: +44 (0)1865 776868
Blackwell Publishing Inc., 350 Main Street, Malden, MA 02148-5020, USA
 Tel: +1 781 388 8250
Blackwell Publishing Asia Pty Ltd, 550 Swanston Street, Carlton, Victoria 3053, Australia
 Tel: +61 (0)3 8359 1011

First published 2006 by Blackwell Publishing Ltd

ISBN-10: 1-4051-3043-1
ISBN-13: 978-1-4051-3043-1

Library of Congress Cataloging-in-Publication Data
 Coutts, John, 1965–
 Loft conversions / John Coutts.–1st ed.
   p. cm.
 Includes index.
 ISBN-13: 978-1-4051-3043-1 (pbk. : alk. paper)
 ISBN-10: 1-4051-3043-1 (pbk. : alk. paper)
 1. Lofts–Remodeling for other use.   I. Title.

 TH3000.L63C68 2006
 728'.314–dc22
 2006000112

A catalogue record for this title is available from the British Library

Set in 10/12.5pt Palatino
by Graphicraft Limited, Hong Kong
Printed and bound in Singapore
by Markono Print Media Pte Ltd

The publisher's policy is to use permanent paper from mills that operate a sustainable forestry policy,
and which has been manufactured from pulp processed using acid-free and elementary chlorine-free
practices. Furthermore, the publisher ensures that the text paper and cover board used have met
acceptable environmental accreditation standards.

For further information on Blackwell Publishing, visit our website:
www.blackwellpublishing.com

Contents

Preface

The purpose of this book is to provide technical guidance on loft conversions in single family dwellings. It is the first book to do so comprehensively and is the result of extensive research and consultation with regulatory bodies and practitioners. It is primarily intended for architects, builders, surveyors and others professionally involved in the process of loft conversion. However, it is also likely to be of interest to students of construction and technically minded householders.

Given the rapid development of loft conversion as a specialist activity in recent years, it is perhaps surprising that so little has been written about the subject. Indeed, some of the construction details illustrated in this book have not been formally described until now.

For clarity and ease of use, the contents of the book are arranged to follow a typical construction sequence. As a matter of necessity, certain sections have been written to reflect guidance and law as it applies in England and Wales.

The regulatory framework within which building work takes place is subject to frequent change. Approved Document guidance for England and Wales, for example, has undergone more than a dozen significant modifications since 2000.

At the time of writing, new versions of Approved Document L *Conservation of fuel and power* and Approved Document F *Ventilation* had just been published. Chapter 10 of this book considers energy conservation in the light of the major changes that have been made both to the Building Regulations and to the Approved Document guidance.

John Coutts
April 2006

Acknowledgements

I would like to thank the following individuals and organisations for their invaluable co-operation: the Office of the Deputy Prime Minister, BRE, Jon Clarke and Peter Guillery at English Heritage, Peter Grimsdale at the Trussed Rafter Association, Celotex, the LFEPA, David Clarke Associates (architects) and LOI Construction. In addition, I would also like to acknowledge the co-operation of Mark Tiddy of Cooper & Turner Ltd for the samples used in the photographs in Chapter 5, TRADA Technology Ltd for the timber grade mark image in Chapter 7 and Polytank Group Ltd for the tank image used in the glossary. Finally, I am indebted to Julia Burden at Blackwell Publishing for commissioning *Loft Conversions*, and to Audrey Tinline for her critical input and encouragement whilst the book was being written.

Material reproduced from the Approved Documents and other government sources is Crown copyright and is reproduced with permission of the Controller of the HMSO.

1 Planning and legal considerations

This chapter examines the influence of planning and other legal mechanisms on the loft conversion process. Obligations imposed by the Building Regulations are considered in Chapter 2.

The controls and mechanisms examined both here and in Chapter 2 are largely separate from each other: planning permission, building control approval and the Party Wall Act, for example, are administered independently. Approvals granted or agreements made under one piece of legislation do not automatically confer rights under another, nor are they intended to. Each element of legislation has specific and generally unrelated aims.

PERMITTED DEVELOPMENT

Most loft conversions in England and Wales are carried out under permitted development, the provisions of which are set out in The Town and Country (General Permitted Development) Order 1995 (the GPDO). The GPDO places specific restrictions on loft conversions by limiting their relative size, form and height. Note that loft conversions in flats and maisonettes are not covered by permitted development and always require planning permission.

Work carried out in accordance with permitted development legislation is effectively granted planning permission by Parliament. There is therefore no requirement either for consultation or for an application to the local planning authority.

The GPDO is based on a presumption that some development *is* permitted and this, in turn, is defined in terms of what is *not* permitted. Classes B and C of Schedule 2, Part 1 of the GPDO deal specifically with alterations and additions to roofs in dwellinghouses. The volume allowances for expansion are reasonably generous and these are based on the type and size of house, and whether it has been extended in the past.

Some of the provisions of the GPDO are subject to interpretation, and this varies considerably between local planning authorities. The following works are often carried out as part of loft conversions, but their status under the GPDO is not clear and interpretation is inconsistent:

- Raising a party wall (see also appeal decision letter in Appendix 3)
- Hip-to-gable conversions
- Providing a highway-facing roof window in a dwelling in a Conservation Area

There are also potential risks when working at the volume limits of permitted development. A local planning authority has discretionary powers to take enforcement action if, in its view, there is an unacceptable breach of planning control. In cases where any degree of doubt exists, therefore, it is prudent to consult the local planning authority before drawings are produced or work commences. The lawfulness of any proposed course of action can be established by applying for a Certificate of Lawfulness (see page 5).

The General Permitted Development Order (England and Wales)

Schedule 2, Part 1, *Development within the curtilage of a dwellinghouse* is relevant to loft conversions. Under Class B, the enlargement of a dwellinghouse consisting of an addition or alteration to its roof is permitted development. However, this is qualified at some length by section B.1:

B.1 Development is *not* permitted by Class B if:

(a) any part of the dwellinghouse would, as a result of the works, exceed the height of the highest part of the existing roof;

(b) any part of the dwellinghouse would, as a result of the works, extend beyond the plane of any existing roof slope which fronts any highway;

(c) it would increase the cubic content of the dwellinghouse by more than 40 cubic metres, in the case of a terraced house, or 50 cubic metres in any other case;

(d) the cubic content of the resulting building would exceed the cubic content of the original dwellinghouse:

    (i)   in the case of a terraced house by more than 50 cubic metres or 10%, whichever is the greater;

    (ii) in any other case, by more than 70 cubic metres or 15%, whichever is the greater; or

    (iii) in any case, by more than 115 cubic metres;

(e) the dwellinghouse is on article 1(5) land.

Under Class C of Schedule 2, Part 1, any other alteration to the roof of a dwellinghouse is permitted development. Class C is qualified by section C.1:

C.1 Development *not* permitted

Development is not permitted by Class C if it would result in a material alteration to the shape of the dwellinghouse.

The meaning of the permitted development provisions

It could be said that permitted development applies to 'normal' buildings in 'ordinary' areas. Unfortunately, there are no better terms than these to describe the majority of buildings and areas to which permitted development rules apply, and the General Permitted Development Order does not offer any; definition is provided only by exceptions, of which there are a considerable number. The GPDO sets out five of them, all related to land designations. Note that, in addition to these exceptions, listed buildings and buildings to which other sorts of planning conditions apply are also excluded from permitted development.

Where permitted development rules apply, and they do for the majority of buildings in England and Wales, it is *not* necessary to apply for planning permission provided the rules described above are adhered to. In order to minimise risk, however, a somewhat conservative approach is sometimes adopted in relation to permitted development, and the rights it confers are not always exploited in full.

The rules of permitted development for loft conversions set out above are largely unequivocal. Nonetheless, they are subject to a degree of interpretation. An explanation of some of the major points is provided below.

Development within the curtilage of a dwellinghouse

This is the title of Part 1 of Schedule 2 of the GPDO. The meaning of curtilage is subject to a degree of interpretation. See Appendix 3.

B.1 (a) Defining the highest part of an existing roof

The ridge (apex) of a roof is generally its highest part. Where the roof is of slated or tiled construction it is usually capped with ridge tiles. But the definition of precisely which part of the ridge is to provide the highest point datum is open to a degree of interpretation, particularly where the original roof is capped with decorative 'crested' ridge tiles. The spine of such tiles generally projects more than 150 mm above the apex.

Similar considerations apply to buildings with party wall parapets (see Fig. 8.1b). In a typical terraced house with a conventional duo-pitch roof, a party wall parapet may extend about 400 mm above the apex. In smaller buildings, or buildings with shallower roof pitches, there would clearly be an incentive to exploit this extra height to increase internal headroom.

In the case of buildings with butterfly roofs and a front parapet wall, local authorities tend to take the roof as the highest point, even though the highway-facing parapet is higher (Fig. 3.1).

As a general rule, therefore, it would be unwise to assume that the top edge of a crested ridge tile or the highest point of a party wall parapet represents the highest point of the roof for the purposes of permitted development. In these circumstances, clarification should be sought from the local planning authority and a Certificate of Lawfulness (Lawful Development Certificate) obtained (see p. 5) before undertaking any work.

B.1 (b) Roof slopes fronting a highway

Alterations to a roof slope fronting any highway (other than the installation of roof windows in the same plane as the existing roof) are *not* allowed under permitted development. For example, a front dormer window (i.e. one occupying and projecting from a roof slope facing a highway) could not be considered permitted development, and planning permission would be necessary.

However, interpretation of the GPDO is less clear cut where hip-to-gable and side-dormer conversions are concerned. Some local planning authorities consider such alterations to be permitted development but this is not the case universally. The local planning authority should be consulted and an application made for a Certificate of Lawfulness.

In the majority of cases, the construction of a loft conversion at the back of a dwelling will satisfy the requirement not to front a highway. However, the GPDO does not differentiate between 'front' and 'back': only the building's relationship with the highway is considered. Inevitably, there are circumstances where the 'back' of a house fronts a highway, for example, where two roads meet at an acute angle or where a house is built on a corner plot. Consultation with the local planning authority is prudent in such cases. It should be noted that the interpretation of 'highway' includes public footpaths as well as roads and lanes.

B.1 (c) Cubic content of roof enlargement

This section of the GPDO defines maximum additional cubic allowances for roof enlargements and alterations. Volume is based on the building's external dimensions. In the case of terraced and semi-detached dwellings, however, it is a matter of interpretation as to whether the full thickness of the party wall, or only half of it, should be taken into account in performing the volume calculation (see the section on curtilage, p. 6).

If the original roof void alone is converted without any projections, enlargement will *not* have occurred as far as permitted development is concerned. Volume allowances for roof enlargements or alterations apply to the *additional* projection from the original roof.

B.1 (d) Original dwellinghouse

'Original dwellinghouse' means the house as it was first built, or as it stood on 1 July 1948 if it was built before then. The volume of the original dwellinghouse includes its roof void. An extension to the original dwellinghouse, if the extension was constructed before July 1948, is not deducted from the volume allowance.

B.1 (e) Article 1 (5) land

The reference to Article 1 (5) land is of importance because it defines where permitted development does not apply and planning permission must be sought. Definition of Article 1 (5) land is provided in Schedule 1 of the GPDO and this constitutes land within:

- A National Park
- An area of outstanding natural beauty
- An area designated as a Conservation Area under section 69 of the Planning (Listed Buildings and Conservation Areas) Act 1990 (designation of conservation areas)
- An area specified by the Secretary of State and the Minister of Agriculture, Fisheries and Food for the purposes of section 41 (3) of the Wildlife and

Countryside Act 1981 (enhancement and protection of the natural beauty and amenity of the countryside)
- The Broads

However, these are not the only exceptions. For example, permitted development rights can also be removed by:

- Article IV directions
- Planning conditions
- Listing

C. Any other alteration to the roof of a dwellinghouse

Class C would allow, for example, the provision of roof windows set into the plane of a highway-facing roof slope, whilst C.1 is something of a catch-all to ensure that the shape of the roof is not altered.

Volume calculation

Volume calculations for planning purposes are based on a building's *external* dimensions (but see *Development within the curtilage* above and *Curtilage: raising party walls* below). When calculating the volume of the original building, porches, conservatories and, of course, the roof space should be included. It *might* also be reasonable to include the volume of any large original chimney stacks. The volume of contiguous outdoor WCs, once a widespread feature of urban terraced houses, might also be considered, perhaps even if these have been demolished.

If the original house has been extended (see above), it will be necessary to calculate the volume of the original house and the volume of the extensions separately in order to determine the cubic allowance for the proposed loft conversion under permitted development.

To determine the volume of the building, the simplest method is to break it down into a number of solid shapes (typically cuboids and triangular prisms) the volumes of which may be calculated and summed to give a total volume (Fig 1.1).

Certificate of Lawfulness

A Certificate of Lawfulness (sometimes Lawful Development Certificate or LDC) is a legal document that may be used to establish that proposed building work is lawful and does not require express planning permission. It is of particular use when working at the limits of permitted development rights.

Application for a Certificate of Lawfulness is made to the local planning authority, which generally bases its decision to grant or refuse on the applicant's submission and drawings. In some cases, the local planning authority will grant a certificate for part of the application. The application should be determined within eight weeks and a fee is payable in all cases.

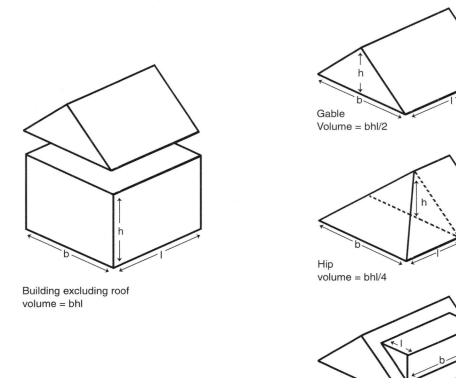

Building excluding roof
volume = bhl

Gable
Volume = bhl/2

Hip
volume = bhl/4

Box dormer
volume = bhl/2

Fig. 1.1   Permitted development volume calculation.

The importance of providing accurate drawings and documentation is emphasised in applying for a Certificate of Lawfulness. Where a certificate is granted, it is of equal importance to ensure that subsequent work conforms to the drawings submitted. Note that section 193 (7) of the 1990 Town and Country Planning Act (as amended by the Planning and Compensation Act 1991) enables the local planning authority to revoke a certificate issued as a result of false information, or if any material information was withheld.

Curtilage: raising party walls and permitted development

In many full-width loft conversions in mid-terrace dwellings, it is practice to raise the full thickness of both party walls to form new flank walls. It is emphasised, however, that a growing number of local authorities consider that such conversions fall *outside* the scope of permitted development because half of each wall is outside the curtilage of the dwellinghouse. Given the degree of uncertainty surrounding the matter, it would be prudent to ascertain the local authority's disposition before undertaking work of this sort. This subject is considered in Appendix 3.

Conservation Areas

Permitted development rights are restricted in Conservation Areas and there is a presumption that any work on a roof in such an area requires planning permission. A new dormer loft conversion, for example, would certainly require an application for planning permission.

Matters are less clear cut in relation to roof lights and roof windows. Many local planning authorities take the view that these, too, require planning permission by virtue of the provisions of the GPDO.

However, some local planning authorities accept the use of co-planar roof windows (i.e. ones that do not project beyond the plane of the roof), even in roof slopes fronting a highway, under permitted development. As a consequence, roof space only conversions are sometimes carried out in Conservation Areas without planning permission.

Because the provisions of the GPDO are open to interpretation, consultation with the local planning authority is advised before undertaking work of *any* description on a roof in a Conservation Area. The section on planning permission below outlines how a planning application may be made and the sources of guidance available.

Article IV directions

An Article IV direction allows the local planning authority to impose additional controls to restrict work that would normally be permitted development. Under an Article IV direction, such work would require planning permission. There is no fee for a planning application that is required as a result of an Article IV direction.

Article IV directions are most commonly used in Conservation Areas, but not exclusively so, and are generally applied to groups of dwellings rather than individual houses. As far as loft conversions are concerned, an Article IV direction might require that planning permission be sought for the installation of roof windows or replacement roofing materials. It should be noted, however, that such work is often considered to require planning permission anyway by virtue of the provisions of the GPDO.

Listed buildings

A loft conversion in a listed building would require both planning permission and listed building consent. Current legislation requires consent to be sought for any works to a listed building that would affect its character as a building of special architectural or historic interest. Note that listed building consent is needed for internal alterations as well as external ones, so even a conversion without projecting elements would require consent.

Listed building consent may also be needed for work on buildings within the grounds of a listed building. It is an offence to carry out any work requiring such consent without first obtaining it.

Applications for listed building consent are subject to similar procedures to planning permission, and the process is administered by the local planning authority. There is no planning fee for an application for listed building consent.

PLANNING PERMISSION

An application for planning permission is made to the local planning authority when a proposal cannot be considered under permitted development. However, before making a formal application and producing detailed drawings, it is prudent to discuss proposals with the local authority's planning department. It is also worth checking local authority planning records to ascertain whether similar conversions have been granted planning permission and to check their planning history. In some architecturally or historically sensitive areas, there is a presumption against *any* roof alterations.

Note that permitted development rights for highway-facing front dormers were removed by the Town and Country Planning General Development Order 1988 (Fig. 1.2).

Fig. 1.2   Front box dormer. Since 1988, conversions of this sort have been excluded from permitted development.

Planning applications

Applications are generally made using a Householder Application form obtained from the local planning authority. Note that this is an application for *full* rather than outline planning permission. Outline permission is generally not sought for routine domestic works such as loft conversions because it is generally possible to ascertain whether a proposal is likely to be granted permission through discussion with the local planning authority.

Once an application has been made, the local authority should decide whether to grant permission (sometimes with conditions) or refuse permission within eight weeks. Where permission is refused, granted with conditions or not decided within eight weeks, there is a right of appeal to the Secretary of State. Where full permission is granted, it is normally valid for three years. Note that the local authority may carry out checks on compliance during and after construction. Some local authorities use their building control surveyors to alert planning enforcement teams to potential transgressions.

Matters such as restrictive covenants, party wall agreements and questions relating to the Building Regulations are not 'material' as far as planning permission is concerned. Similarly, moral issues, such as the personal circumstances of the applicant, are not normally taken into account.

An increasing number of local authorities are accepting electronic applications made using the Government's Planning Portal. This allows users to submit and pay for applications online. This does not affect the way planning decisions are made. Whether submitted electronically or on paper, most of the questions on the Householder Application form are relatively straightforward. However, attention should be given to the following points:

- Ownership
- Location plan
- Site layout plan
- Floor plans and elevations
- Drawings (general)

Ownership

The application must include a completed certificate relating to the ownership of the land. If the proposed conversion involves raising a party wall, the statement of ownership must reflect that the property to which the application relates is not entirely owned by the applicant and there is a requirement to formally notify other owners about the application before it is submitted. Note that this is not the same as a Party Wall Agreement.

Location plan

An up-to-date plan at 1:1250 or 1:2500 scale is required that accurately shows the property in relation to roads and other properties. A large-scale Ordnance Survey extract is generally used in this application, although map details contained in deeds may also be acceptable. The application site (generally the house and garden) must be outlined in red. Any other land owned by the applicant in the vicinity is outlined in blue.

Site layout plan

The site layout plan (1:500 or larger scale) should indicate the position of the existing property, site boundaries, adjoining houses (including extensions to them), outbuilding (including sheds), existing or proposed parking spaces, trees and hard surfaced areas including patios and paths.

Floor plans and elevations

Plans are drawn to a scale of either 1:50 or 1:100 and should distinguish between existing and proposed structures. Floor plans should show the room layout for the whole building with one drawing for each floor. Doors, windows and wall thicknesses should be indicated. Elevations should show what the proposed conversion will look like from the outside. Where neither side of the conversion is visible, a section drawing should be provided. The drawings must indicate the building materials to be used. Photographs showing aspects of the existing site may be of assistance in making an application.

Drawings general

All plans and drawings should be to a recognised scale and only metric measurements should be used. Plans should be clearly identified and drawing numbers provided. Locator and site layout plans require a north point. Note that there is no restriction on the size of drawings. Whilst it is convention for architects and building designers to produce large-format drawings, some local planning authorities prefer A3 drawings which are easier to duplicate.

Note that the ease with which drawings can now be reduced or enlarged has increased the risk of scaling errors: for example, an original 1:100 drawing at A1 becomes 1:200 if the sheet is reduced to A3 (and only approximately 1:200 if the reduction is carried out optically). Care should therefore be taken when drawings are resized, or when scaling off drawings that are not original.

PLANNING GUIDANCE SPECIFIC TO LOFT CONVERSIONS

Some local planning authorities produced guidance specific to loft conversions and roof alterations in specific localities. When producing drawings for an application for planning permission, the following sources may be considered in support of a planning application:

Unitary Development Plan (UDP)

This is the development plan produced by certain unitary district authorities and London boroughs. The policies in the UDP have special status afforded by the Town and Country Planning Act 1990 in deciding planning applications. The UDP may set out acceptable roof alterations in some detail.

Supplementary planning guidance (SPG)

Many local planning authorities produce supplementary planning guidance that expands on their statutory policies. Some local authorities produced highly detailed supplementary planning guidance on loft conversions. The contents of such guidance, whilst non-statutory, are taken into account as a material planning consideration when determining applications. The weight accorded to SPG increases if it is prepared in consultation with the public and has been the subject of a council resolution.

Design guides

Because much planning law is essentially negative, local planning authorities are increasingly producing local design guides to clarify what *can* be done. These guides are usually drawn up for Conservation Areas, often after consultation with residents. In the case of loft conversions, a design guide might provide indications of acceptable scale, and use of materials in a proposed dormer construction. Where adopted, design guides have the same status as supplementary planning guidance and are thus taken into account as a material planning consideration.

OTHER CONDITIONS AFFECTING DEVELOPMENT

Planning conditions affecting permitted development

'Planning conditions' (there is no other specific technical name for them) are used to remove permitted development rights and they are being used increasingly to restrict development such as loft conversions. Planning conditions are most often applied to new high density housing, although not exclusively so. Where a local planning authority has applied such a condition, it is necessary to apply for planning permission. Some local authorities, however, waive the normal application fee in such cases.

Because planning conditions may apply to areas and building types that would traditionally have enjoyed permitted development rights, property owners are sometimes unaware of their existence. Conditions such as these ('charges' in legal terms) are recorded in the Local Land Charges register which local authorities are required to maintain. There is, therefore, a case for checking the register even in an area where a building could reasonably be assumed to have permitted development rights.

Restrictive covenants

A restrictive covenant imposes conditions on how an owner may use land. This sometimes includes alterations and extensions to buildings. In cases where a restrictive covenant applies, the permission of the original developer may be required before an extension can be built. The restrictive covenant should appear in the land register entries to the title of the property. Note that the register referred to in such cases is maintained by the Land Registry and is quite separate from the Local Land Charges register.

A restrictive covenant can take many forms and can apply to more or less any type of property. It should be noted that local authorities widely apply such covenants to council houses sold under right-to-buy rules and that the local authority therefore has the benefit of the covenant. However, it is generally possible to escape the covenant through negotiation. In cases where there is a very old and apparently out-of-date restrictive covenant, it is sometimes possible to insure against the risk of enforcement. It is stressed that restrictive covenants are generally a civil matter and operate independently of the planning system.

Mortgage lenders

Where property is mortgaged, it may be necessary for the householder to obtain bank or building society permission before undertaking a loft conversion. This is because any work might affect the lender's interest in the building and would apply whether or not it had advanced the money to pay for the work. Some lenders may charge a fee for consent.

Buildings and contents insurance

It is a requirement of most household insurance policies to inform the insurer before undertaking any building alterations and to adjust premiums accordingly. This applies to both buildings and contents insurance. In addition, it is generally necessary to modify policies to reflect the greater size of the property once work is complete.

Tree preservation orders

A tree preservation order (TPO) is an order made by the local authority to protect a tree or group of trees. An application must be made to the local authority to fell or undertake work, including pruning, on a tree subject to a TPO. Such matters are generally dealt with by the local authority's planning department.

Bats

Bats and their roosts are protected under the Wildlife and Countryside Act 1981 and the Conservation (Natural Habitats &c.) Regulations 1994. It is an offence to remove or disturb bats without notifying the relevant Statutory Nature Conservation Organisation (SNCO) before undertaking any work. In the United

Kingdom, these are English Nature, the Countryside Council for Wales, Scottish Natural Heritage and, in Northern Ireland, the Department of the Environment.

UNIFIED CONSENT REGIME

At the time of writing, the government is conducting a review of the processes that govern some commonly conducted home improvements such as loft conversions. The Householder Development Consents Review will consider, amongst other things, the possibility of combining consents in previously independently administered areas such as planning and building control approval. In theory at least, this could pave the way for a unified consent regime.

THE PARTY WALL ETC. ACT 1996

The Act usually applies when lofts are converted and it provides a framework for preventing and resolving disputes in relation to party walls. Anyone planning to carry out work of the kinds described in the Act must give notice of their intentions to the adjoining owner. The Act applies in England and Wales and is invoked by the building owner. Local authorities are not usually involved in the process.

In semi-detached and terraced dwellings, it is usually necessary to carry out work on a wall shared with another property (a party wall) as part of a loft conversion. Any work on a party wall, other than minor operations, is likely to fall within the scope of the Party Wall etc. Act 1996. Some common examples of work on party walls as part of a loft conversion include:

- Cutting into an existing party wall to provide support for a beam
- Constructing a dormer stud cheek over part of an existing party wall
- Raising the height of an existing party wall to form a new flank gable wall
- Raising a compartment wall if there is no separation in the roof void

Section 2 of the Act deals with work to existing party walls and, if the correct procedures are followed, it confers a number of valuable rights, including rights of access (subject to conditions). It is emphasised that the Act should be invoked even if the proposed work does *not* extend beyond the centreline of the party wall. For example, the Act would still apply where a beam with a bearing of 100 mm is to be supported by a 225 mm wall.

Perhaps the single most important factor in preventing disputes from arising is the building owner's relationship with the adjoining owner. Wherever possible, the building owner should discuss the proposed work with the adjoining owner before notice is served.

Procedure

The building owner must serve a *party structure notice* on any adjoining owner. A party structure notice must be served at least two months before the planned starting date for work to the party wall. Note that the notice is only valid for one year. There is no official form for giving notice, but the Act stipulates that the notice contains the following:

- The name and address of the building owner (joint owners must all be named)
- The nature and particulars of the proposed work (drawings may be included)
- The date on which the proposed work will begin

Although not specifically mentioned in the Act, it might be prudent also to include in the notice:

- The address of the building to be worked on
- The date of the notice
- A statement that it is a notice under the provisions of the Party Wall etc. Act 1996

The adjoining owner's agreement and written consent to the proposed work, if it is received, does not relieve the building owner of obligations under the Act – for example, the requirement to avoid unnecessary inconvenience whilst work is carried out. Neither does it eliminate the possibility of differences arising between the adjoining owner and the building owner once work has begun. The two-month period between serving notice and the planned commencement of work may be reduced with the agreement of the adjoining owner.

Disputes

If the adjoining owner disagrees and does not consent to the work proposed, one of two approaches may be adopted:

- The building owner and adjoining owner concur in the appointment of a single surveyor – the 'agreed surveyor'. The agreed surveyor produces an 'award' (sometimes called a party wall award) that sets out in detail the work proposed and conditions attached to it.
- The building owner and adjoining owner each appoint a surveyor to settle any differences and to produce an award as outlined above. In such cases, the appointed surveyors select an additional surveyor – the 'third surveyor' – who would be called in to mediate if the two appointed surveyors cannot reach an agreement.

Note that if the adjoining owner fails to respond to the party structure notice within 14 days of service, a dispute is considered to have arisen. In this case, a surveyor is appointed on behalf of the adjoining owner and the procedure described above is followed.

The Act defines a surveyor as a person who is not party to the matter. This means, in theory at least, that anybody can act as a surveyor in a party wall dispute. In practice, of course, it would be prudent to appoint, or agree on the appointment of, a person with a good knowledge both of construction and of administering the Act. Surveyors appointed under the dispute resolution procedure of the Act must consider the interests and rights of both the building owner and the adjoining owner. They do not act as advocates for the respective owners, and an award must be drawn up impartially. Fees for the surveyor (or surveyors) are generally paid by the building owner.

Details of surveyors specialising in the Party Wall Act may be obtained from the Royal Institution of Chartered Surveyors or the Pyramus & Thisbe Club. Contact information is provided in the bibliography.

2 The Building Regulations and building control

The meaning of 'building work' is set out in Regulation 3 (1) of the Building Regulations 2000 (as amended) for England and Wales. The definition is a broad one and, with few exceptions, a loft conversion – and certainly one containing habitable rooms – falls within it. By virtue of this, a loft conversion must comply with the applicable requirements of Schedule 1 of the Building Regulations 2000 (as amended). It should be noted that the Building Regulations are law, not guidance; powers to enforce the regulations are contained in the Building Act 1984. The penalty for non-compliance includes a fine of up to £5000 plus the requirement to pull down or alter any work that is carried out in contravention of the regulations.

The Building Act 1984

The Building Act 1984 is the enabling Act under which Building Regulations are made. It provides the broad legal framework for building control and gives the Secretary of State the power to make Building Regulations.

The Building Regulations

These are sometimes referred to as the Principal Regulations. Note that the regulations are subject to frequent amendments. In the 18 months from April 2002, seven sets of amendments were published. At the time of writing, a consolidated version is not available.

Schedule 1 of the Building Regulations 2000 (as amended) sets out a number of functional requirements with brief explanations. No technical detail is provided. With the possible exception of Part M, all are potentially relevant to loft conversion:

- Part A   Structure
- Part B   Fire safety
- Part C   Site preparation and resistance to contaminants and moisture
- Part D   Toxic substances
- Part E   Resistance to the passage of sound
- Part F   Ventilation
- Part G   Hygiene
- Part H   Drainage and waste disposal

- Part J   Combustion appliances and fuel storage systems
- Part K   Protection from falling, collision and impact
- Part L   Conservation of fuel and power
- Part M   Access to and use of buildings
- Part N   Glazing – safety in relation to impact, opening and cleaning
- Part P   Electrical safety

Approved Document guidance

The Approved Documents, so called because they are approved and issued by the Secretary of State, provide technical guidance that supports Schedule 1 of the Building Regulations. There are Approved Documents to support each of the 14 parts of Schedule 1 listed above, and an additional document to support Regulation 7 – *Materials and workmanship*. In tandem with the Building Regulations they support, the Approved Documents are frequently updated.

It must be emphasised that each of the Approved Documents is intended only to address the requirements of the regulation it supports. Whilst each is internally consistent, it should not be assumed that it meshes seamlessly with guidance offered in other Approved Documents. In addition, many expressions used in the Approved Documents have meanings that apply only within those documents. Even common terms should not be taken at face value. For example, in Approved Document A, the word 'wall' is used in its accepted perpendicular sense while in Approved Documents B and C, 'wall' can also mean a part of a roof pitched at more than 70 degrees to the horizontal.

Approved Document guidance offers a number of valuable concessions without which many conversions would not be possible. These include:

- *Approved Document B – Fire safety (2000 as amended)*
  — Use of an enclosed stairway and egress window rather than a fully protected stairway (this guidance is currently under review – see Chapter 4)
- *Approved Document K – Protection from falling, collision and impact (1992)*
  — Reduced headroom for stair access to lofts
  — Use of alternating-tread stairs in certain cases
- *Approved Document L – Conservation of fuel and power (2006 edition)*
  — Insulation requirement may be reduced where it would affect floor area (but see Chapter 10)

Relationship between the Building Regulations and the Approved Documents

Building Regulations and the Approved Documents are not the same thing, although they are often confused. The Approved Documents are often wrongly described as 'building regs'. In simple terms, the Building Regulations are law and the Approved Documents are guidance. The use of Approved Document guidance is not mandatory provided that work complies with the Building Regulations.

In practice, however, the relationship between the Building Regulations and the Approved Documents is rather more complex. The Approved Documents are an important instrument for the implementation of government policy and, as such, there is an increasing emphasis on quantified standards. Some of the Approved Documents are prescriptive in character and in some cases (Approved Document L, for example) it would be misleading to suggest that only guidance is offered because specific targets are set out.

There are also important legal presumptions associated with the use of Approved Document guidance. Section 7 of the Building Act 1984 states that if in any proceedings, civil or criminal, it is alleged that a person has contravened a provision of the Building Regulations, then a failure to comply with an Approved Document may be relied upon as tending to establish liability. Equally, proof of compliance with the Approved Documents will tend towards negative liability.

BUILDING CONTROL

There is a legal obligation to use a building control service when carrying out building work. The purpose of building control is to provide an independent check that building work complies with the Building Regulations (but see also *Statutory notifications*, p. 21). Checks on compliance are chiefly achieved through site inspections but, depending on the procedure chosen, plans may also be inspected before construction commences. Building control services are offered by local authorities and also by private approved inspectors. In either case there is a charge for the service. It is important to realise that building control and planning are separate, independently administered functions. Even though a loft conversion may not require planning approval, it will be subject to Building Regulations if habitable rooms are to be built.

Local authority building control

Local authority building control is provided by district or borough councils in England and Wales. It is invoked either through a full plans application or under a building notice. There is generally no cost difference between the two procedures and either may be used. Because loft conversions are seldom straightforward, some local authorities specifically recommend that they be carried out under the full plans procedure.

Full plans

Drawings and other relevant construction details are deposited with the local authority before work commences. The application is checked and the local authority must make a decision within five weeks. This may be extended to two months by agreement. Plans may be:

- Approved
- Approved subject to conditions
- Rejected

In all cases, it is necessary to inform building control of commencement at least two working days before work starts. It is possible to start work before a decision on the application is made, provided that a commencement notice is issued, but the protection offered by full plans does not become active unless a notice of approval is issued.

The advantage with full plans approval is that it reduces uncertainty about the technical specification of the work before it starts. The cost of a full plans application is generally commensurate with that of a building notice.

Under full plans, it is possible to seek a determination from the Secretary of State if the plans, or any part of them, are rejected. The disadvantage with the full plans procedure is that the process can take some time.

Full plans application

The full plans application form provided by local authorities is intended only to elicit the most basic information including the site address as well as the purpose and description of the proposed works. It is therefore incumbent upon the person carrying out the work to provide as much detail as possible in order to indicate compliance with *all* the relevant Building Regulations. This information is usually provided in drawings, plans, specification sheets and structural engineering calculations.

Whilst there is no such thing as a 'model' full plans application, attention to the following points (some of which are set out in regulation 14 of the Building Regulations) is recommended.

Block plan: this must be to a scale not less than 1:1250 showing the size and position of the building and boundaries of the site. This is important in relation to fire safety because it allows an assessment to be made about potential external access for rescue purposes, ladder access for egress windows and possible unprotected areas in relation to relevant boundaries.

Plans and sections: these should be to a scale not less than 1:100 and preferably 1:50 and provided not just for the conversion itself but for every other floor in the building. Both existing and proposed elements of the building should be included. This is of great importance in a loft conversion where the layout and structure of even remote elements of the building can have a considerable impact on the design and indeed the feasibility of the conversion, particularly with regard to fire safety and structural stability.

Structural engineering: calculations justifying the dimensions, weight, position, fixings and bearings of any beam either timber or steel. In a typical loft conversion, calculations would generally be required for the following elements:

- Steel floor beams
- Roof beams

- Trimmers (floor and roof)
- Header beams over windows
- Spine wall, if proposed for new load-bearing use

Electrical safety: some local authorities require a statement indicating that work is to be designed, installed, inspected and tested in accordance with the requirements of BS7671 and Building Regulations Part P (*Electrical safety*). An indication of whether or not the work is to be conducted under the Competent Persons Scheme may also be requested.

If the plans comply with the Building Regulations, the local authority will issue a notice stating that the plans have been approved. Note that if work does not commence within three years, a new application must be made.

Where the local authority is not satisfied with the application, it may ask for amendments to be made or additional details to be provided. Alternatively, it may issue conditional approval. This will either specify modifications which must be made to the plans, or it will specify further plans that must be deposited. The local authority may only apply conditions if the person carrying out the work has either requested them to do so or has consented to them doing so. A request or consent must be made in writing. If plans are rejected the reasons are stated in the notice.

Common problems with full plans applications

Approval of plans may be withheld for a number of reasons in the case of a loft conversion, either because detail is inadequate or missing, or the proposal appears to contravene the regulations. Some of the more common reasons are listed below:

- Dimensions of structural members such as floor and roof beams with supporting engineering calculations absent or inadequate (Part A)
- Fire safety details including failure to provide an adequately protected stairway (Part B)
- Incorrect position and/or size of emergency egress windows (Part B)
- Staircase and landing details with particular attention to headroom (Part K)
- Insufficient thermal insulation (Part L)
- Details of electrical works inadequate or statements absent (Part P)

Of the points listed above, the provision of an adequately protected stairway and the availability of headroom on staircases/landings are potentially the most difficult to remedy.

Pre-application prior to full plans application

A number of local authorities now operate a plan checking service which may be used *before* a full plans application is made. A charge is made for this service. Comments made or advice given by local authorities under these circumstances does not constitute formal approval.

Building notice

The alternative to making a full plans application is to conduct work under a building notice. Less detail is required and the information needed in the notice is set out in regulation 13. It includes:

- A description of the building work and its location
- 1:1250 scale plan indicating the size, position and curtilage of the building and its relationship to adjoining boundaries
- Width and position of any street on or within the curtilage of the building
- Number of storeys
- Drainage provisions
- Insulation details where a cavity wall is to be filled
- Details of hot water storage system (if this is to be provided)
- Details of how electrical works are to be carried out

Work may commence within two clear working days of the notice being submitted to the local authority. A building notice ceases to have effect three years after it is given to the local authority, unless building work commences before that time has elapsed. The cost of making a building notice submission is generally the same as that for a full plans application. Like building work carried out under the full plans procedure, construction conducted under a building notice is subject to site inspections.

The disadvantage with the building notice procedure from a builder's perspective is that it is not possible to ascertain in advance whether the proposed work complies with the Building Regulations because detailed plans are not submitted.

Obligations under building notice procedure

Using a building notice does not remove the obligation to provide whatever information is necessary to demonstrate compliance with the Building Regulations (regulation 13 (5)). Many local authorities insist on plans and sections of the whole building in order to demonstrate compliance with fire safety and headroom requirements. In addition, structural engineering calculations are nearly always required for loft conversions. To quote from the building notice submission form provided by one local authority:

> 'Structural calculations and details may be required to justify elements of work carried out. As they are always required for a loft conversion, you are advised . . . to consider the use of the full plans application form instead of this building notice.'

Notification and inspection of work

Statutory notifications

Whether the work is conducted under a local authority building notice or a full plans application, the Building Regulations impose an obligation on the person carrying out the work to provide notification before and after certain stages. Interestingly, the primary obligation is for the person carrying out the work to notify, rather than for the local authority to inspect. Notification periods are set out in regulation 15 and they are:

■ Before commencement of work (two days)
■ Before covering up any excavation for a foundation (one day)
■ Before covering up any foundation (one day)
■ Before covering up any damp proof course (one day)
■ Before covering up any concrete or material laid over site (one day)
■ Before covering up any drain or sewer to which the regulations apply (one day)
■ After laying, haunching or covering any drain or sewer in respect of which Part H of Schedule 1 imposes a requirement (notice to be given no more than five days after completion of work)
■ Where a building, or part of one, is to be occupied before completion (notice to be given at least five days before occupation)
■ After completion of building work (notice to be given no more than five days after completion)

Site inspections

As noted above, the Building Regulations 2000 (as amended) sets out requirements for notification rather than inspection. In practice, of course, site inspections are carried out to ensure compliance with the Building Regulations. However, it should be noted that under regulation 18 of the Building (Amendment) Regulations 2001, the local authority is explicitly permitted to carry out tests on more or less *any* relevant aspect of building work in order to establish compliance.

Whilst there is no formal list of inspections (other than those implied in *Statutory notifications* above), the following is provided as a 'model' inspection schedule. Some of the following inspections may, of course, be combined:

■ Commencement.
■ Steel beams in loft area and any raising of chimney stack.
■ Floor, wall and roof timbers. Location of escape windows.
■ Insulation to roof and walls including ventilation of voids. First fix electrical.
■ Drainage of sanitary fixtures. Extractor fans and other ventilation arrangements.
■ Fire doors, self-closers, smoke detectors and fire alarms. Other fire resistance measures. Stairs, headroom, handrails and room ventilation.
■ Completion.

Loft conversions are generally built at considerable speed and it is important to ensure that work is not covered up before inspection. The order of inspections must reflect the sequence in which work would be concealed.

Certificate of completion of work

Under a full plans application, the local authority may issue a certificate of completion of work provided that it is satisfied that the work complies with the

Building Regulations and that it was requested to do so when the plans were initially submitted. In the case of a building notice, the local authority is not required to issue a completion certificate, although most will. A completion certificate does not provide conclusive evidence of compliance. However, it is a legal document and its significance should not be underestimated: note that a completion certificate is normally required by a purchaser's solicitor when a house is sold.

Disputes

Most minor disagreements that arise between persons carrying out building work and the building control service over whether or not plans comply with the requirements of the Building Regulations are settled by discussion and, where necessary, alterations to works. However, where it is not possible to reach an accommodation, the Building Act 1984 provides a number of procedures for resolving disputes. These include:

The determination procedure (under full plans only)

Where the local authority or approved inspector believes that plans do not comply with one or more of the Building Regulations, but the person carrying out the building work contends that they do, it is possible to seek a determination from the Office of the Deputy Prime Minister in England or, if in Wales, the National Assembly for Wales. In general, the application must relate to work that has not substantially commenced.

Under this procedure, which varies depending on which building control system is used, the arguments of both parties are taken into consideration and a decision (determination) is made as to whether the applicant's proposals comply with the Building Regulations. For example, an applicant may consider that the provision of a sprinkler system in lieu of an enclosed staircase would comply with Requirement B1 *Means of warning and escape*. The local authority's case might be that a fire-resisting enclosure is required. The Secretary of State considers both sets of arguments and makes a determination. Appendix 2, *Appeals and determinations* contains some examples. Note that the determination procedure is not available for building work carried out under a building notice – it only applies where a full plans application has been made.

Between 1998 and 2004, determinations were sought in ten cases involving loft conversions in single-family dwellings, most of them concerned with B1 *Means of warning and escape*. In all but one of the examples, the Secretary of State concluded that proposals did not comply with the Building Regulations.

The relaxation/dispensation procedure (full plans or building notice)

The Building Act 1984 gives the power to local authorities to relax or dispense with a particular requirement in Schedule 1 of the Building Regulations. It is therefore possible to apply to the local authority to seek a *relaxation* or a *dispensation* of one or more requirements. An application to relax or dispense with a requirement may be made under either full plans or building notice and may be applied equally to both planned and completed building work. If an approved private inspector is being used, an application for relaxation or dispensation is made to the local authority.

There is a critical difference between the determination procedure and the relaxation/dispensation mechanism. From the applicant's point of view, the determination procedure is based on the premise that a proposal *does* meet the requirements of the Building Regulations. But under the relaxation/dispensation procedure (which is administered by local authorities rather than central government), the applicant must recognise that a proposal *does not* comply with a particular requirement, or part of one.

Under the *relaxation* procedure, it is necessary to state the reasons why a particular *part* of a requirement in Schedule 1 is believed to be too onerous or inappropriate and give reasons for not complying with it. Where it is proposed not to comply with the whole of a requirement, a *dispensation* is sought.

Note that many of the requirements of Schedule 1 of the Building Regulations are life safety matters and that local authorities are unlikely grant either relaxations or dispensations in practice.

Appeals

If a local authority refuses an application for relaxation or dispensation, there is a right of appeal under section 39 of the Building Act 1984 against that decision. Appeals (in England) are made to the Office of the Deputy Prime Minister or, if in Wales, to the National Assembly for Wales. An appeal must be made within one month of the date of being notified of a refusal.

The basis of any such appeal must be an acceptance by the appellant that proposed (or completed) building work does *not* comply with regulations – what is sought is a relaxation or dispensation of the regulations. This is still widely misunderstood, and an analysis of the arguments advanced in appeal proceedings indicates that some appellants are under the impression that the work they are proposing complies in some way.

In cases relevant to loft conversions in single-family dwellinghouses, a total of ten appeals were made between 1998 and 2004, six seeking relaxations/dispensations concerning Requirement B1 *Means of warning and escape* and four regarding Requirement K1 *Stairs, ladders and ramps*. All were dismissed. It is perhaps interesting to note that in two cases concerning Requirement K1, the Secretary of State considered that the work in question complied with the regulations anyway. See Appendix 2 *Appeals and determinations* for decisions concerned with loft conversions.

Buildings of historic or architectural interest

The Building Regulations now take greater account of historic buildings and new provisions are set out in regulation 9 of the Building and Approved Inspectors (Amendment) Regulations 2006. Some Approved Document guidance also reflects the need for flexibility.

A more flexible approach is generally applied to alterations rather than extensions to such buildings. It would be hard to envisage a situation where a reduced standard of fire safety would be acceptable in a domestic loft conversion. A more realistic consideration, perhaps, might be the proposal of a new dormer window with a reduced standard of thermal insulation in a conservation area loft conversion. In this case, it would be prudent first to establish through discussion, and then to determine through a full plans application, that such a proposal was acceptable.

Electronic building control applications

This facility, known as Submit-a-Plan, allows applicants or agents to submit building control applications online. Where the system is supported by the relevant local authority, the following applications may be made:

- Full plans submission
- Building notice
- Regularisation certificate
- Replacement doors and windows

Submissions made in this way have the same status as those made under the traditional hard copy system. An electronic application has the obvious advantage of being nearly instantaneous and it reduces the need to produce paper copies of everything. In the case of a building notice, however, it is still necessary to provide notice two working days before commencing work.

Approved inspector building control service

Whilst using a building control service is a legal requirement, there is no obligation to use the service provided by the local authority; a private approved inspector may be used instead. At the time of writing, there are about 50 companies and individuals offering an independent building control service, although not all of them provide a service for small-scale works such as loft conversions. Fees for the Approved Inspector service are agreed by negotiation, there is no prescribed scale, and these are generally competitive with those charged by local authorities.

The approved inspector procedure operates slightly differently from local authority building control. The person carrying out the building work and the approved inspector must first jointly issue an *initial notice* notifying the local authority of the intended building work. Once this notice has been accepted by the local authority, the responsibility for plan checking and inspection is placed in the hands of the approved inspector.

If required, the approved inspector will issue a plans certificate confirming that the proposed building work complies with the Building Regulations. On completion of the work, the approved inspector issues a *final certificate* to the local authority to indicate that work described in the initial notice is complete. Where work does not comply with the Building Regulations and agreement on remedial action cannot be reached with the approved inspector, the inspector may cancel the initial notice. In such cases, the building control function would revert to the local authority.

Approved inspectors are subject to procedures set out in the Building (Approved Inspectors etc.) Regulations 2000 as amended. However, the functional regulations administered (i.e. the Building Regulations 2000 as amended) are of course the same for both local authority and private approved inspectors.

3 External forms

This chapter considers different types of conversions and the factors that influence their choice. A number of basic forms are described. These are simplified somewhat for the sake of clarity and it is emphasised that hybrid forms incorporating several different elements are common. Drainage is considered in this chapter because it has a significant but sometimes overlooked impact on the appearance of the conversion: many otherwise reasonable designs are spoiled through lack of attention to the external arrangement of discharge stacks, branches, vents and rainwater goods.

Note that a proposal to alter the roof may require planning permission. All habitable conversions must comply with the Building Regulations. In the majority of cases it is necessary also to invoke the Party Wall etc. Act (1996) before undertaking work. Planning and other regulatory requirements concerning loft conversions in England and Wales are covered in detail in Chapters 1 and 2.

Dormers

'Dormer' describes a projecting vertical window, with its own roof, that is set into a principal roof slope. The word entered the English language in the late sixteenth century, although the form itself predates this. Historically, dormer windows were subordinate structures intended to provide light and ventilation to the roof space in order to make it habitable. In contemporary usage, however, the word dormer has become associated with more or less any habitable projection from a roof, regardless of its scale, and it is in this more recent sense that the word dormer is used here.

INFLUENCES ON FORM

A loft conversion is never a blank slate. Physical and regulatory factors play a major role in determining form, and perhaps a larger one in the case of loft conversions than in other sorts of small-scale building work. In some respects, these constraints simplify the design process but only in the sense that they provide absolute limitations.

Pitch, plan and headroom

The relationship between the pitch of the existing roof slopes and the plan of the building determines the availability of headroom in the conversion. The finished floor to ceiling height of the main portion should generally be 2.3 m or more.

Where headroom (other than at the ridge) is restricted, adopting a box dormer design may be the only viable approach.

Stair access

Even where a pitched roof space provides sufficient headroom without any apparent need for dormer projections, it is still essential to make a separate assessment of headroom on the new stairway. Consideration should be given to roof slopes or downstand elements (such as purlins) that are likely to foul the stair. In some cases, for example where the existing roof is hipped, it may be necessary to construct a side dormer to provide adequate headroom.

Shallow pitched roofs

In roofs with a relatively shallow pitch (32.5 degrees or less), a rear box dormer is the most effective means of providing daylight and headroom, although it is not necessarily the most attractive form. Note that the effectiveness of traditional small dormer forms in admitting daylight is considerably reduced in low-pitched roofs: as pitch decreases, cheek length increases to form a dormer 'tunnel' (cheeks may be glazed but see *Design considerations* below). Traditional dormers work better and are generally more pleasing to the eye where the main roof slope has a pitch greater than 40 degrees.

Roof form – gable

Roofs with gable walls, including the pitched roofs of terraced dwellings with compartment walls that extend to ridge height, are generally easier to convert than hipped roofs. Apart from providing a greater internal volume for a comparable area, gable walls are often capable of providing support for new elements of structure such as floor and ridge beams.

Roof form – hip

In an unmodified state, hipped roofs are generally more troublesome to convert than gabled roofs. The hip or hips restrict the useable plan area of the conversion. Hips also make the task of supporting new ridge and floor beams more difficult. In addition, it should be noted that in many cases, an unmodified hipped roof is not capable of providing adequate headroom over the proposed stairway if this rises beside an external wall. Where it is not possible to accommodate the conversion or elements of it within the existing hipped roof configuration, construction of a side dormer or full hip-to-gable conversion may be considered.

Roof form – butterfly or London roof

This form is typically encountered in terraced dwellings of the nineteenth and late eighteenth centuries (Fig. 3.1). It has a distinctive V-shaped profile with a central

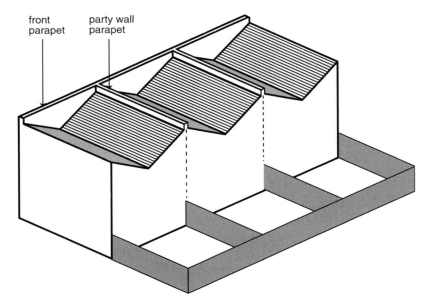

front
parapet

party wall
parapet

Fig. 3.1 Butterfly roof.

gutter running at right angles to the front of the building. For the sake of uniformity in terrace construction, the twin slopes of the butterfly roof are generally hidden from view behind a parapet wall at the front of the building and are usually only visible from the rear. In effect, the roof comprises a pair of independent lean-to slopes. The compartment walls separating such dwellings generally project beyond the roof covering to form party wall parapets. The roof slopes themselves are frequently pitched at a very shallow angle: slopes of less than 20 degrees are common. The two roof voids thus formed are generally too small to justify conversion in their own right.

Because there is not a great deal of 'loft' to convert in these cases, it is generally necessary remove the roof slopes entirely to accommodate a new floor and roof structure. Typically, a mansard solution is adopted (Fig. 3.2e). The party wall parapets on each side are built up to form flank walls while to the front elevation, a mansard slope is created behind the parapet wall, with rainwater collection provided by a box gutter running between the two. To the rear, the V-shaped gap between each half-gable is filled in with brick or studwork and raised to full height. A flat roof is provided, generally with party wall parapets configured to project above it.

Note that a conversion of this sort always requires planning permission because it alters the shape, height and visibility of the roof. In addition, it is emphasised that the guidance in Approved Document B specific to loft conversions relates only to the conversion of an *existing* roof space (see Chapter 4). It is not generally applicable where new habitable rooms are created by the replacement of the whole of an existing roof with a mansard roof structure.

Drainage

The provision of bathroom and WC facilities must be carefully considered because the drainage of these will have a significant impact on both the internal arrangement and the external appearance of the conversion. In practical terms, the relatively broad vertical façades of most box dormers generally make the fixing of external drainage a relatively straightforward matter. By contrast, accommodating branch, discharge and vent pipes on or near the relatively narrow face of a traditional gabled dormer can only be achieved with difficulty and is likely to detract from any aesthetic gains.

Constructing the dormer face in the same plane as the existing rear wall may simplify the process of connecting the new system to an existing discharge stack. It also eliminates the often unsightly arrangements that arise when rainwater pipes and soil/vent pipes have to be offset to cross the boundary between the old and new parts of the building. Note, however, that building up the dormer face in the same plane as the wall beneath it will tend to create an overwhelming structure. Consideration should be given to altering the position of the discharge stack relative to the conversion at a point lower down the building, rather than at loft level, to simplify connection.

Within the conversion, sanitary pipework should generally be kept above floor level in order to avoid conflict between pipes and elements of structure at or below floor level. It is therefore good practice to finalise the position of appliances at an early stage in the design process.

Where possible, sanitary appliances should be positioned as near as possible to the external discharge stack to limit the length of branch pipes and thus ensure that pipework is able to penetrate the external wall above the level of the finished floor or other obstructions such as floor beams. The right-hand column of Table 3.1 indicates the relative height of the underside of a branch pipe to the floor at its maximum length. Cutting through beams to accommodate pipework is generally not feasible. Where floor joists are underslung (see Fig. 7.1), the supporting beam will be above relative floor level and may foul the passage of pipes passing through the external wall.

It is sometimes not possible to achieve the required slope for pipework for appliances with low-level drainage outlets such as WCs, baths and shower trays, particularly over extended runs. Where this is the case, appliances may be

Table 3.1   Unvented branch connections – appliance to wall/floor junction.

Appliance	Typical branch pipe diameter (mm)	Minimum slope (mm/m)	Appliance to stack (max)	Height of branch pipe above floor level at max length (approx)
WC	100	18	6 m	+ 32 mm
Bath/shower	50	18–90	4 m	− 22 mm
Bath/shower	40	18–90	3 m	− 4 mm
Hand basin	50	18–90	4 m	+ 378 mm
Hand basin	40	18–90	3 m	+ 396 mm
Hand basin	32	20 (at 1.75 m)	1.75 m	+ 415 mm

mounted on a dais to achieve the required fall. Commercially available electric macerator units capable of pumping foul water are an acceptable alternative where conventional gravity drainage is not possible.

Chimneys

In many cases, it is necessary to incorporate a chimney stack within the design of the conversion. Where the stack is at the ridge of the roof, this is generally a straightforward matter. However, where the stack rises from the eaves on a rear elevation, it will be necessary either to remove it (if it is no longer in use) or build around it. Where the stack is retained and remains in use, it will often be necessary to increase its height to conform to guidance set out in Approved Document J *Combustion appliances and fuel storage systems*. The dormer structure may be built to encompass the stack. Note, however, it is generally not permissible to fix structural elements to chimney stacks or breasts (see also Fig. 7.15).

Planning considerations

The influence that planning legislation has on the form of the conversion depends on the building itself and also the area it is in. Further details are included in Chapter 1. A synopsis is provided below:

- *Permitted development*: limited restrictions on loft form provided that rules on height, shape, volume and the relationship between the conversion and highway are observed.
- *Conservation Areas/Article IV* – conversions altering roof shape require planning permission. Provision of roof windows may also require planning permission. Local guidance on acceptable dormer forms may be available.
- *Listed buildings* – planning permission and listed building consent required for any alterations, including internal modifications.

Emergency egress windows

Note that the layout of rooms within the conversion may be influenced by external factors. Rooms requiring emergency egress windows, for example, must be accessible by ladder in most cases. The presence of a conservatory or carport beneath such a window would generally represent an unacceptable obstacle as far as safe ladder access was concerned (see also Chapter 4).

CONVERSION FORMS

To a certain extent, roof modifications defy rigid classification. However, a number of common basic configurations are described below. Some of these are traditional 'whole roof' forms (such as the mansard) while others are roof adaptations (such as the hip-to-gable conversion). For the sake of clarity, traditional subordinate dormer forms are considered separately in the next section.

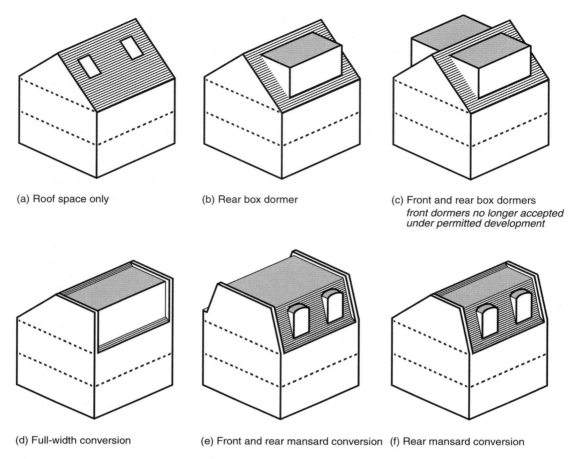

(a) Roof space only   (b) Rear box dormer   (c) Front and rear box dormers
                                                          *front dormers no longer accepted
                                                          under permitted development*

(d) Full-width conversion   (e) Front and rear mansard conversion   (f) Rear mansard conversion

Fig. 3.2   Conversion forms.

Roof space only conversion

This describes a conversion that exploits the void bounded by the existing roof slopes (Fig. 3.2a). It is sometimes also described as a roof light, attic or VELUX conversion. Appropriate structural modifications are carried out in order to create habitable accommodation, but these do not result in alterations to the external shape of the roof. Daylighting is provided by roof windows or roof lights set into the roof slope rather than dormer projections. This approach is generally only suitable where the roof space is of an adequate size, with headroom available over a reasonable floor area.

Box dormer conversion

The box dormer is now the most popular form of loft conversion in conventional pitched roof dwellings (Fig. 3.2b). It is capable of providing a relatively large amount of useable space quickly and at a low cost. It is also generally the least satisfactory form from an aesthetic perspective. Box dormers are generally built on the opposite side of the building to the highway to qualify as permitted development and for this reason they are sometimes described as rear dormers.

In a typical configuration, the conversion's rear elevation (the dormer face) rises vertically at or near the eaves and in many cases may occupy a large proportion of the width of the roof. The triangular sides of the dormer (the cheeks) rake into the existing roof slope. Both face and cheek walls are generally constructed from timber studwork with tile, slate or sometimes timber cladding. Occasionally sheet metal coverings are used. A flat roof spans from below the ridge of the existing roof to the dormer face. The flat roof is generally laid to fall in the direction of the face to simplify rainwater drainage and collection via guttering. Note that, in small dormers, the roof may be configured with a reverse fall for the sake of appearance (see *flat dormer* below).

The provision of a flat roof provides an extended area of full-height ceiling within the conversion. This is particularly advantageous in conversions of small/medium dwellings with a relatively shallow roof pitch where useable headroom, and therefore floor space, would otherwise be restricted to the area immediately beneath the ridge.

The cut roof structure of many small to medium-sized dwellings (up to 7.5 m between wall plates) is not well adapted to the use of traditional small dormers. There is generally not enough headroom to accommodate pitched roof dormer forms and even where there is, the sill position is generally fouled by a single purlin which must be removed. Given the complexity of adapting the roof for relatively little increase in volume, it is generally considered easier, and less expensive, to remove the entire rear roof slope. Note that in larger cut-roof dwellings (those where the plan and pitch generate *slope* lengths in excess of 6 m), there are generally two purlins per slope, and it is sometimes possible to accommodate a traditional dormer window between these.

Front box dormer conversion

Front box dormers gained a considerable degree of notoriety during the 1980s and, largely on aesthetic grounds, highway-facing dormers were removed from permitted development in 1988 (Fig. 3.2c). Any projection beyond the plane of a highway-facing roof slope now requires planning permission and this is generally denied where front box dormers are proposed.

Full width dormer with masonry flanks

This form is similar to the rear box dormer and offers similar advantages in terms of extending useable floor space (Fig. 3.2d). The principal difference is that the dormer occupies the full width of the roof with flank walls providing a direct vertical continuation of the gable or party walls beneath: the conversion is thus bounded by masonry flank walls rather than tile hung cheeks. As with the box dormer, the dormer face is vertical and generally clad with tile or slate. This approach is best suited to terraced dwellings that are separated by party wall parapets: compartment walls of this sort project above the roof slope and walls may be raised without the necessity of disturbing the roof covering of the adjoining property.

Raising party wall parapets to form new flank gable walls maximises space within the conversion and introduces a degree of vertical discipline to the building's rear elevation. However, because the face of the dormer and its supporting masonry flanks are closely aligned with the existing rear wall, such a conversion will tend to overwhelm the building beneath. Note that this form of conversion is sometimes, incorrectly, described as a mansard conversion (see below).

Mansard conversion

The defining quality of the traditional mansard roof is its broken slope, with a long, steep lower section and a shorter and flatter upper slope. This form takes its name from the French architect François Mansart (1598–1666), although he did not devise it. In classical architecture, the mansard's double slope rises from the eaves and is applied to each elevation of the building. When applied only to opposing elevations, it is sometimes described as a gambrel or gabled mansard.

The term 'mansard' is now applied rather vaguely to any roof with a steeply pitched principal slope. In the case of a mansard loft conversion, a steeply-pitched lower slope is applied to the rear elevation and sometimes to the front as well. Where it is applied to an urban front elevation (Fig. 3.2e), it may rise from behind a parapet wall as an adaptation of a butterfly roof. Because the modern form often has a flat upper slope, it is sometimes called a half mansard.

The mansard is a recessive form, in the sense that it is inclined, and it is therefore effective in reducing the apparent mass of a converted roof space. Properly detailed and appropriately applied, the mansard is perhaps the most pleasing of the wide loft forms. In a number of inner-London boroughs, adopting a mansard-style slope is sometimes a planning requirement on highway-facing elevations when terraced roofs are adapted (see *Roof form – butterfly* above). Note that an alteration of this sort is generally considered to constitute a roof extension rather than a loft conversion for the purposes of planning and building control.

Mansard conversions are sometimes applied to the rear elevation only, with gables or compartment walls extended upwards to form flank gables (Fig. 3.2f). As with full-width dormers with masonry flanks (above), this approach is most easily adopted in terraced dwellings where the existing party wall parapets project above the roof slope. For the sake of appearance, the new flank gable walls are generally raked back to match the pitch angle of the mansard. The mansard slope, which is usually inset from the masonry flanks, is generally clad with slate or tile. Lead, copper or zinc cladding may also be used.

Daylighting within the conversion is provided either by vertical windows (which by virtue of the sloping mansard face will be dormers in their own right), by roof windows set in the same plane as the mansard slope, or by recessed windows. In contemporary practice, the upper roof slope is generally configured as a conventional flat roof laid to fall towards the face; guttering and fascia are, of course, not provided at the junction between the slopes. Care should be taken in detailing the junction. A lead flashing fixed beneath the upper slope roof covering, dressed over and clipped to the principal slope, is, for example, preferable to a felt downstand.

In order to preserve unity, particularly in terraces where other conversions are anticipated or have already been carried out, local authorities will generally specify the angle of pitch on highway-facing slopes (alterations to the roof slope other than the introduction of roof windows always require planning permission). Typically, a slope angle of 70 degrees to the horizontal is required. Note that there are a number of technical requirements associated with roof pitch: part of a roof pitched at an angle of more than 70 degrees to the horizontal to which persons have access is defined as an external *wall* rather than a roof for the purposes of Approved Document B *Fire safety*. A similar pitch-based distinction is recognised in Approved Document C *Site preparation and resistance to contaminants and moisture*, although this document indicates that its moisture resistance provisions apply to roofs of any pitch.

Hip-to-gable conversion

The hip-to-gable conversion, in which an existing pitched roof hip is replaced by a new vertical gable wall, usually at the side of the building, is sometimes carried out as a conversion in its own right in order to enlarge an existing roof space (Fig. 3.3b). Generally, though, hip-to-gable conversions are carried out in order not only to increase the volume available, but also to provide support for new structural members such as ridge and floor beams, and sometimes new purlins, as part of a larger dormer conversion. In addition, such a conversion may be

(a) Terrace end - hipped roof   (b) Hip-to-gable   (c) Side dormer

Fig. 3.3 Hip treatments.

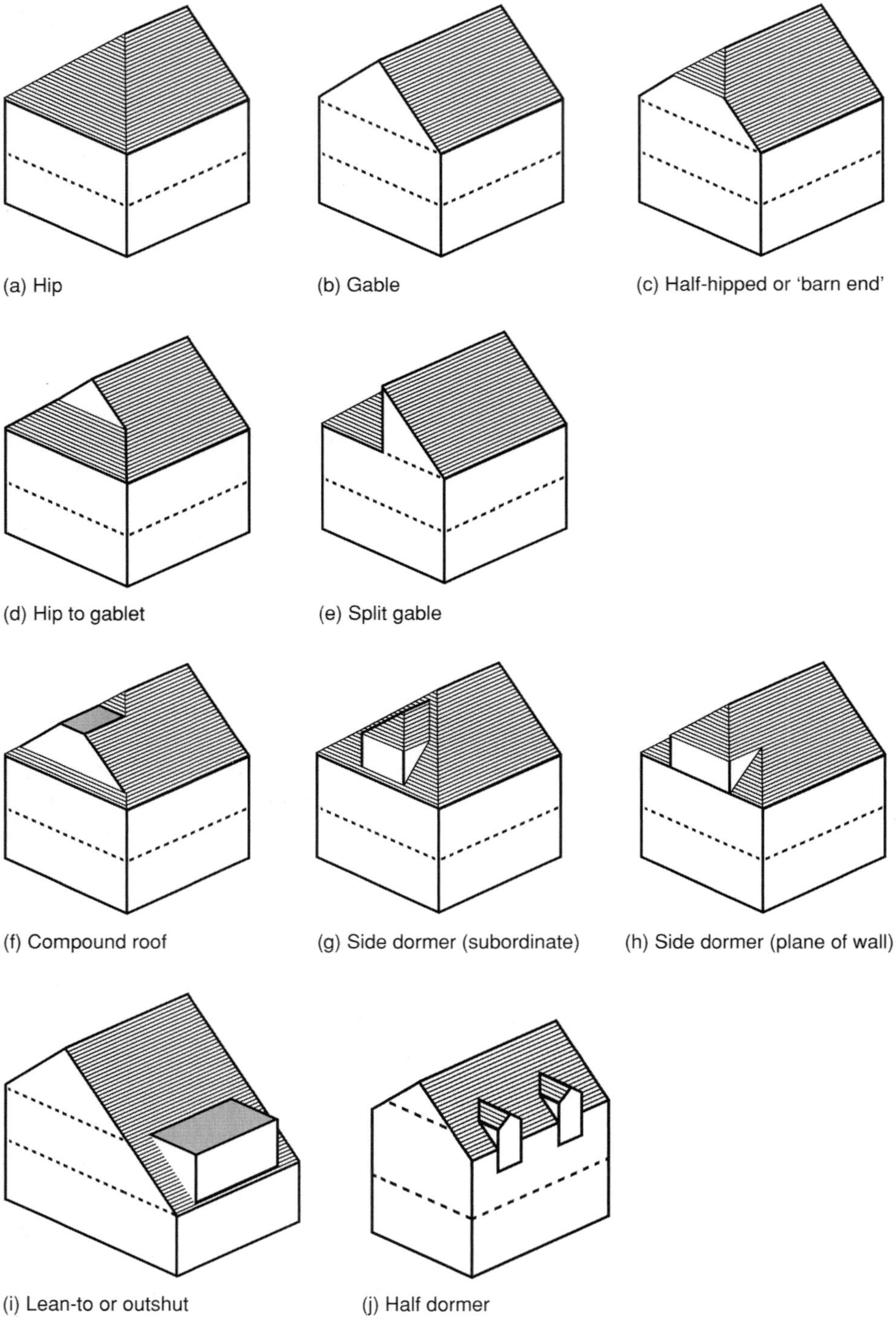

(a) Hip

(b) Gable

(c) Half-hipped or 'barn end'

(d) Hip to gablet

(e) Split gable

(f) Compound roof

(g) Side dormer (subordinate)

(h) Side dormer (plane of wall)

(i) Lean-to or outshut

(j) Half dormer

Fig. 3.4   Roof and gable treatments.

necessary to provide headroom for a new staircase. A new gable end may be formed in two ways:

- *Masonry* – the gable end is built up in brick to match the existing. Alternatively, blockwork may be used and given a render finish.
- *Studwork* – the gable end is built up in timber studwork. The stud structure may be battened out for tile hanging. Alternatively, expanded metal lathing may be fixed and a render finish applied. Appropriately designed, a timber studwork gable is capable of providing support for a steel ridge beam.

Note that a number of variant hip roof treatments are possible that combine both gable and roof elements (Figs 3.4c, d, e and f).

Side dormer conversion

The side dormer may either be set within the roof slope or built up on the head of the gable wall (Fig. 3.3c and Figs 3.4g and h). Like the hip-to-gable conversion described above, it replaces part of an existing roof hip. Side dormers are sometimes adopted where a full hip-to-gable conversion is not acceptable for planning reasons and are often built to provide headroom clearance for a new stairway. Where built off the gable wall, they may also serve to accommodate and conceal the ends of floor beams that might otherwise project beyond the plane of a roof slope.

Lean-to conversion

Some dwellings have a rear single-storey ground level outshut accommodated beneath the roof slope (Fig. 3.4i). Projections of this sort were sometimes created intentionally at the time of construction, but are often the result of subsequent extension. Exploitation of the void formed by the roof at first floor level is potentially rather more straightforward than a conventional conversion, if only because there is generally no need to provide an additional staircase. In a contemporary context, a suitable outshut might be formed by a garage at the side of a dwelling. It is emphasised that the floor of a loft conversion or any other room formed over a garage must provide full (rather than modified) 30-minute fire resistance (Approved Document B *Fire safety* 9.5 and 9.14).

Half dormer

This term is used to describe a dormer window that occupies the junction between an external wall and the roof, for example, in the upper floor of a $1^{1}/_{2}$ storey dwelling where rooms are formed partly within the walls and partly within the roof space of the house (Fig. 3.4j). It is not a 'pure' external dormer form in a strict sense: most of the traditional forms considered presently in this chapter may be configured as half dormers. Because it rises directly from the wall below, the face of a half dormer is generally constructed from masonry rather than studwork.

Existing attic rooms

Slate was rapidly adopted as an urban roofing material from the 1790s. One of the consequences of this, initially, was lower roof pitches (often less than 30 degrees) and therefore roof voids that were generally unusable for habitation. However, steeper pitches became fashionable from about 1860 onwards and certainly in larger dwellings (those with roofs spanning in excess of 8 m) this led to the creation of substantial roof voids that could be exploited. Attic rooms, with roof lights and dormer windows to provide light and ventilation, are thus original elements in many substantially proportioned Victorian houses.

A degree of confusion surrounds the regulatory status of original habitable attic rooms. In some cases, purpose-built Victorian attic rooms might satisfy, or perhaps even exceed, current structural guidance, but that is all: in their 'as-built' form, it is most unlikely that they would meet other current standards. There is no requirement that they do, *unless*:

■ The purpose to which the building is put is changed (a *material change of use*, see Glossary).

■ Material alterations are undertaken. For example, the installation of a new WC/bathroom would have to conform to current standards in an 'as-built' attic. However, there would generally not be a requirement to upgrade anything else, even though it may be prudent to do so. Regulation 3 of the Building Regulations 2000 (as amended) recognises that even if a building does not comply with existing regulatory requirements, any alterations undertaken should not make those elements 'more unsatisfactory' (but see also Chapter 10 on renovation of thermal elements).

Sleeping gallery

Where a roof void is open to the rest of the building, a raised floor may be constructed to exploit part of the roof space. A sleeping gallery (sometimes called a *sleeping loft* or a *half loft*) is thus created, usually only at one end of the building. Access is provided by stairs or, under some circumstances, by a fixed ladder with handrails. Guarding is required at the gallery edge and the height of the gallery should not exceed 4.5 m above ground level (see also Fig. 4.16 and Appendix 2).

Floor platforms of this sort are suited to certain types of single-storey buildings, including some barn conversions. In Wales, a sleeping gallery formed in the roof void is called a *crog lofft* and it is a traditional feature in some small rural dwellings, particularly in west Wales. Although of some antiquity (conversions of this sort are probably as old as the pitched roof itself) the form remains popular.

TRADITIONAL DORMER FORMS

The primary function of the dormer window is the provision of light and ventilation in the roof space (Fig. 3.5). Traditional dormer forms are generally

(a) Gab

Fig. 3.5 Traditional roof adaptation. The modest proportions of these half dormers and the relatively steep roof pitch contribute to a pleasing whole.

subordinate to the overall roof structure. Correctly proportioned and appropriately positioned, they may enhance the appearance of the building.

Traditionally, the relatively low mass of dormer windows meant that they often required support only from the rafters and trimming members. However, it is emphasised that the depth of structure needed to accommodate insulation material and the requirement to provide double glazing means that modern recreations of such traditional forms result in structures of greater mass. Additional means of support for the dormer (including support by the floor) must be taken into account.

Careful attention should be given to the choice of cladding and roofing material: large format slates and tiles, unless they are a feature of the principal roof, often appear clumsy on small dormers. This applies in particular to the more complex hipped and eyebrow roof forms. Other cladding and roofing materials that may be considered include lead, zinc, copper, timber and GRP. Equally, expanded metal lathing may be fixed to the vertical studwork, and render applied.

Dormer cheeks may be glazed, rather than clad, in order to improve daylighting. Where this is proposed, the questions of overlooking and the practical difficulties of cleaning the external face of the glazing panels, which are generally fixed, must be addressed.

Traditional dormer forms are generally described by roof rather than window shape. Note that the roof of the dormer, particularly if it is gabled, is sometimes described as the *dormer head*. The following are 'pure' rather than hybrid forms:

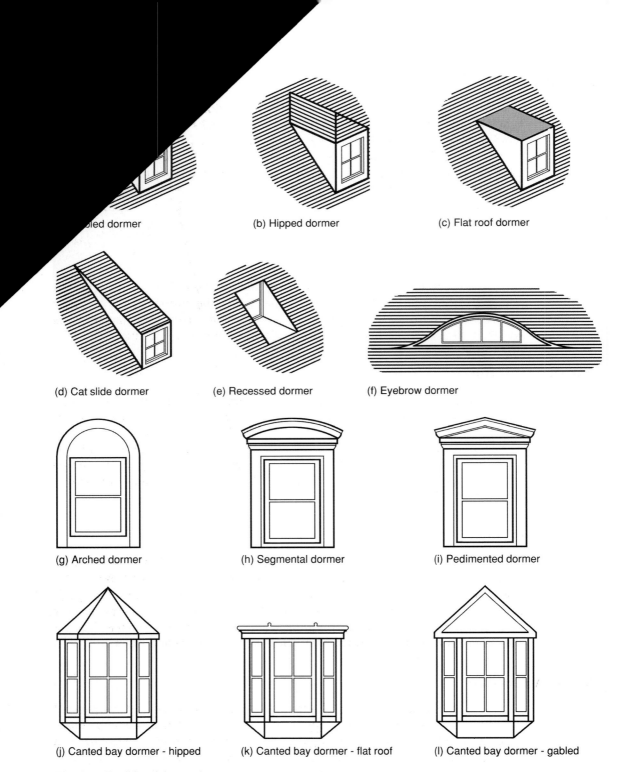

(b) Hipped dormer

(c) Flat roof dormer

...led dormer

(d) Cat slide dormer

(e) Recessed dormer

(f) Eyebrow dormer

(g) Arched dormer

(h) Segmental dormer

(i) Pedimented dormer

(j) Canted bay dormer - hipped

(k) Canted bay dormer - flat roof

(l) Canted bay dormer - gabled

Fig. 3.6 Traditional dormer forms.

Gabled dormer

This is perhaps the most widespread of the traditional dormer forms, with a symmetrical duo-pitched roof, vertical studwork cheeks and a casement or sash window to the face (Fig. 3.6a). Valley detailing is required at the junction between dormer and principal roof slopes. Depending on headroom, the ceiling within the dormer itself may be fixed directly to the rafters or to ceiling joists spanning the dormer slopes. Gabled dormers are sometimes described as *pitched*, *cottage* or *bonnet* dormers. The pitch angle for the dormer and principal roof are normally the same.

Hipped dormer

As above, but with a hip end rather than a gable (Fig. 3.6b). This form works best where the principal roof itself is hipped. The pitch of the dormer slopes normally matches that of the principal roof. The term *piended* (pronounced peended) is widely used in Scotland to describe a hipped roof. See also Canted bay.

Flat dormer (small)

Structurally the most straightforward of the dormer forms because there are no pitched roof components and therefore valley boards/valley rafters are not required (Fig. 3.6c). The flat roof may be finished with a covering of lead, copper or zinc; in the case of lead, roll detailing may be a prominent feature. A built-up felt covering may also be used. In order to eliminate the need for guttering, which might appear clumsy on a small dormer, the flat roof may be configured with a reverse fall to the principal roof slope.

Cat slide dormer

This form is most effective on steeply-pitched roofs (Fig. 3.6d). In effect, the cat slide dormer is a monopitched outshut of the main roof, with acute triangular cheeks, that falls to the face of the dormer at a shallower angle. Because the pitch of the dormer slope may be relatively shallow, it is important to ensure that the roofing material is capable of providing weather and uplift resistance. The cat slide is also sometimes described as an *eased-out*, *wedge* or *swept* dormer.

Recessed dormer

The recessed dormer is sometimes adopted where planning restrictions prevent projections from the roof slope, or perhaps where a small balcony is to be introduced (Fig. 3.6e). As its name suggests, the vertical window is encompassed *within* the envelope of the roof. For practical reasons, including daylighting, rainwater drainage and the impact on available floor area, this solution is best suited to roofs with a relatively steep pitch. It may be used to accommodate windows within a mansard slope. It is sometimes called an *inset* dormer.

Eyebrow dormer

The eyebrow dormer has no sides, the roofing material rising and falling over the opening as an unbroken undulation (Fig. 3.6f). This form allows for the creation of a broad but relatively shallow rectangular window opening at the face. In carpentry terms, it is the most complex of all the dormer forms. When designing an eyebrow window, care should be taken to ensure that the roofing material is capable of conforming both to the slope and to the curvature of the projection.

Arched dormer

This has a true semi-circular roof profile: the commensurate circle would have a diameter to match the width of the dormer (Fig. 3.6g). For practical reasons, these generally require metal cladding and, for the sake of appearance, this is often continuous across cheeks and roof. It is sometimes described as a *barrel* dormer.

Segmental dormer

Segmental profiles are generally gently curved, with an arc struck from a point below the springing line (Fig. 3.6h). The roof is generally finished with lead, copper or zinc. It represents a classical approach and is generally associated with front elevations. Sometimes called a *bow* dormer.

Pedimented dormer

The pedimented dormer is similar in some respects to the gabled dormer, at least in the sense that it is triangular in profile (Fig. 3.6i). However, like the segmental dormer described above, it is a classical feature and somewhat formal. The pitch of the pedimented dormer is generally shallow, in the Grecian style, and the roof is generally finished with a metal covering rather than slate or tile. It is often associated with steeply pitched roofs including mansards, and enlivened with painted moulding.

Canted bay dormer

Sometimes described as a *polygonal* or *bay dormer*. The front of the dormer is three-sided, with a central window flanked by glazed areas that are angled back. The hipped (piended) form (Fig. 3.6j) is widespread in Scotland. Figs 3.6k and 3.6l illustrate flat and gabled versions. The gabled form tends to top heaviness. Canted bays provide good light admittance relative to dormer length.

DESIGN CONSIDERATIONS

Most loft conversions are not designed by architects, and relatively little attention is given to external form. In most cases the primary intention of the building owner is to create the greatest possible volume of habitable space at the least possible cost and an unfortunate consequence of this is that many conversions are carried out with scant regard for basic architectural principles.

Untrammelled by aesthetic considerations, the suburban box dormer has thus evolved as a 'pure' form to occupy the regulatory niche provided by Schedule 2 of the General Permitted Development Order. To this extent, most loft conversions design themselves, with concrete plain tiles, off-the-shelf PVC-U windows, plastic guttering and deep fascia boards completing the visual envelope. In this sense, the box dormer has emerged not only as a pure form, but a distinctive and instantly recognisable one as well. But that is not to say that it is necessarily 'good' architecture: indeed, it is not architecture at all, and the worst box dormers are contrivances of extraordinary ugliness. It is not always a question of relative size: the reason many loft conversions fail aesthetically is simply that they bear no relation to the buildings beneath them.

In order to tie old and new together, the conversion should at least echo elements of the existing building, and an appreciation of the existing structure is therefore essential. Attention is drawn to the following points:

■ Fenestration
■ Roof details
■ Vertical cladding and roofing
■ External drainage

Fenestration

New windows should respect the basic proportions of openings in the existing building. Equally, the proportions of the glazed units and thickness of glazing bars *within* individual windows should be considered in relation to those existing in lower floor windows.

The arrangement of windows should also be considered. In many cases, vertically aligning new windows with existing windows is appropriate. Where this is not possible, creating a number of evenly-spaced openings is generally better than an asymmetrical window arrangement, particularly if this bears little relationship to a sequence of openings lower down the building.

In order to reduce the somewhat unrelieved appearance of the dormer face, windows may be set back into reveals, but the depth of the dormer skeleton must be configured to allow this (e.g. the use of 150 rather than 100 mm studwork). Windows should not be in the same plane as the wall unless other openings in the building are configured in this way.

Roof details

Both flat and pitched roofs are considered more pleasing to the eye where they oversail the walls beneath them to some extent. In the case of a typical box dormer (where a flat roof projection of about 100 mm provides visual separation between wall and roof) this is generally practical only to the rear elevation and not the cheeks. Fascia boards should be relatively shallow and painted a dark colour; white fascia boards simply draw attention the gutter. Guttering should be returned against the dormer cheeks at each end.

Mansard roofs tend to make single-storey buildings appear top heavy and they are only aesthetically satisfactory in existing buildings of two or more storeys. Box dormers, of course, have the same effect. Nevertheless, inclining the face of the dormer in a conversion at any level serves to reduce its impact to some extent.

Vertical cladding and roofing

The selection of cladding material should reflect the existing roof material, but note that large format slates and tiles are often visually cumbersome and are generally considered unattractive as vertical cladding materials. Note also that, for practical purposes, large format tiles, particularly profiled tiles, are not suited to complex detail formation, such as hips, on small dormer roofs. Plain tiles (265 × 165 mm) are the most widely available small-format covering and are often used for cladding conversions.

Lead flashings and soakers should be treated with patination oil at the time of fixing to reduce lead carbonate staining on the cladding below them.

The pitch of new roof elements, such as gabled dormers, need not match the pitch of the principal roof exactly, unless an element of the principal roof (such as a cross gable) presents itself in the same plane as the dormer. Hipped dormers reduce the apparent mass of the dormer structure and help it to merge with the roof slope. For visual balance, and to simplify construction, all slopes of a hipped roof are pitched at the same angle. Note that a roof pitch of 45 degrees is considered unattractive.

New brickwork for built-up gables should match the existing masonry. Brick type, mortar composition and bonding must be considered.

Glazed cheeks allow for increased light levels within the conversion, and fixed side glazing of this sort is sometimes used in small dormers. When evaluating this approach, the following points should be considered: fire resistance – relationship between any glazed cheek and a boundary; thermal insulation – the effects of increasing the glazed area; cleaning – providing safe access to allow cleaning of glazed side panels.

External drainage

Good designs are often let down by ramshackle external drainage. The position of external drainage elements should be considered at an early stage in the design process. Consideration may be given to re-routing discharge stacks lower down the building and not at roof level where complex changes of direction are likely to be conspicuous.

4 Fire

Providing an appropriate means of escape from fire for loft conversions is potentially complex. This chapter considers means of escape and other elements of fire safety in loft conversions in single-family dwellinghouses. As with other aspects of the conversion process, it is not possible to adopt a one-size-fits-all approach to fire safety: each conversion must be assessed individually.

It is likely that a number of significant changes to the guidance provided in Approved Document B Fire safety *will be introduced during 2006. If made, these changes will have an impact on the process of loft conversion, and some of the measures set out in this chapter will no longer apply. The proposed changes are outlined at the end of this chapter.*

Regulatory framework

Part B of Schedule 1 to the Building Regulations 2000 is in five parts, all of which interact to some degree:

- B1 Means of warning and escape
- B2 Internal fire spread (linings)
- B3 Internal fire spread (structure)
- B4 External fire spread
- B5 Access and facilities for the fire service

All are relevant to loft conversions. Current guidance supporting the requirements is contained in Approved Document B (AD B) *Fire safety* (2000 edition consolidated with 2000 and 2002 amendments). Additional guidance relevant to stairways is provided in Approved Document K *Protection from falling, collision and impact* (1998 edition, amended 2000).

There is no obligation to follow guidance provided in any of the Approved Documents, provided that it can be demonstrated that a proposal fulfils the requirements of the Building Regulations in some other way. However, it should be noted that AD B *Fire safety* offers considerably more practical guidance than other Approved Documents. This guidance is widely used and, as far as loft conversions are concerned, adhering to it is often found to be less time consuming than proving compliance through other methods (see Appendix 2 for examples of this).

FIRE RESISTANCE: BASIC REQUIREMENTS

Satisfying the guidance concerned with internal and external fire spread is generally a straightforward matter and compliance may be achieved through the use of

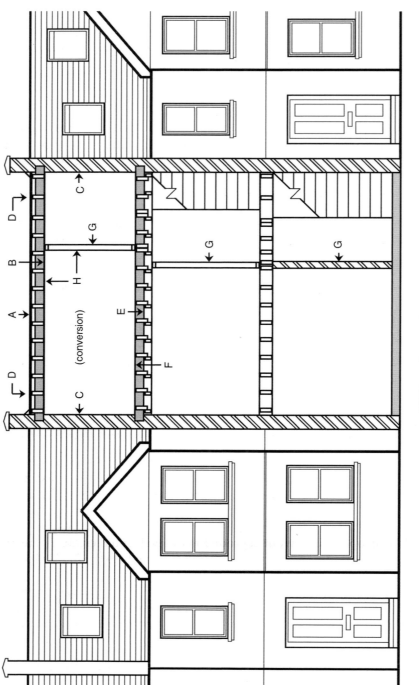

KEY (see also notes in main text):

A - roof covering   F - new floor
B - roof beam   G - internal walls, partitions and
C - compartment walls   elements of structure
D - dormer cheeks (to rear)   H - spread of flame (linings)
E - floor beams

Fig. 4.1   Loft conversion fire resistance: basic requirements.

appropriate fire-resisting construction. Fig. 4.1 illustrates the basic standards of fire resistance required in a single-family dwelling where the converted roof space creates a three-storey building.

- *A – Roof covering*. A bitumen felt roof covering (irrespective of felt specification) is deemed to be designation AA and therefore acceptable if:
    (i) the deck is a minimum of 6 mm ply and
    (ii) a finish of bitumen-bedded stone chippings covering the whole surface of the roof for a depth of at least 12.5 mm is provided.
- *B – Roof beam*. A roof beam is excluded from the definition of an element of structure and need not be protected unless it is essential for the stability of an external wall which needs to have fire resistance, or the roof is part of an escape route.
- *C – Compartment walls*. Walls separating buildings must have a minimum period of fire resistance of 60 minutes.
- *D – Dormer cheeks*. See Fig. 8.8 for fire resistance requirement.
- *E – Floor beam*. This is an element of structure and requires 30-minute fire resistance.
- *F – New floor*. 30-minute fire resistance.
- *G – Internal walls, partitions and elements of structure*. These require 30-minute fire resistance.
- *H – Spread of flame*. The use of conventional plasterboard and plaster finishes will generally satisfy the requirement to inhibit internal fire spread.

Note that additional requirements for specific types of conversions are covered in subsequent drawings.

WARNING AND ESCAPE

Satisfying the requirement to provide appropriate means of escape from a loft conversion is sometimes a complex task. The main reason for this is that the relationship between the escape route(s) and rooms throughout the *entire* dwelling must be taken into account. For example, a proposed loft conversion that would lead to the creation of a three-storey dwelling in which the only stairway discharges into an open-plan layout at ground floor level would create a potentially intractable problem, unless the staircase and escape route to the final exit were to be enclosed.

Guidance on means of escape and the provision of alarm systems is provided in Approved Document B *Fire safety* (B1 *Means of warning and escape*). Although necessarily complex, the guidance is, in essence, underpinned by two quite simple principles:

- *Early warning* – a linked automatic fire detection and alarm system must be provided for each storey when a loft is converted.
- *Unassisted escape* – occupants should be able to escape from the conversion to a safe place without outside assistance.

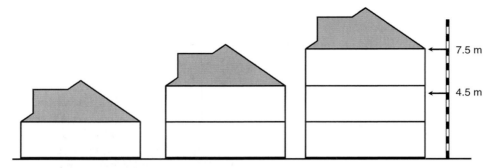

Fig. 4.2   Critical floor heights.

The principle of unassisted escape is specifically relaxed to some extent in the case of loft conversions that form a three-storey dwelling, but only under certain tightly defined circumstances.

The risks presented by fire in a lower-storey to the occupants of upper storeys increases considerably with the height of the building. Equally, increasing height has an effect on the type of escape that is possible. Thus, in broad terms, the primary escape dependency for dwellings in Approved Document B is floor height (see also Fig. 4.2):

- *Floors less than 4.5 m above ground level*: provision of egress window(s) for self-rescue.
- *Floors 4.5 m or more above ground level*: provision of EITHER a protected stairway OR an enclosed stairway and egress window(s) for ladder rescue (but only under certain circumstances), OR an alternative escape route to a final exit.
- *Floors at or above 7.5 m above ground level*: provision of BOTH a protected stairway AND an alternative escape route.

The number and form of escape routes required, and the level of protection that must be provided for them, is considered in the following sections. It should also be noted that, in all cases, the guidance is generally *less* onerous where the dwelling has more than one stairway serving the conversion, provided that the stairways provide alternative means of escape and are adequately separated from each other. Whilst there are a number of caveats and additional requirements, the primary objective is to provide the occupants with a stairway that leads all the way from the loft conversion to a safe final exit (such as a front door) and does not pass through any habitable rooms (such as an open-plan living area).

Floor height rules

Rules for floor height measurement are referenced in AD B 2.7 (preamble) and set out in Appendix C to the Approved Document. The Approved Document drawing is reproduced here with modifications (Fig. 4.3). While assumptions are often made about the likely height of top floors based on the number of storeys in the building, specific site conditions (for example, sloping ground or basement

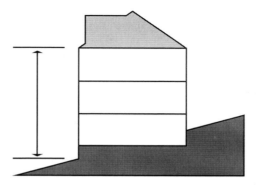

Fig. 4.3 Height of top storey.

'areas') can have an impact on this. Note that the height measurement is referenced to ground level on the *lowest* side of the building.

Storey numbering rules

There is no generally accepted storey designation system in the UK. The top storey of a three-storey building, for example, may variously be described as the 'second' or 'third' storey.

Note that Approved Document B uses an ordinal storey-numbering system based on floor designations, i.e. the storey at ground level is the ground storey (not the first storey), the next storey is the first storey (not the second storey) and so on (Fig. 4.4). When refering to the whole building, the Approved Document uses a cardinal system based on absolute storey numbers; thus a 'two-storey house' is one with just a ground storey and a first storey.

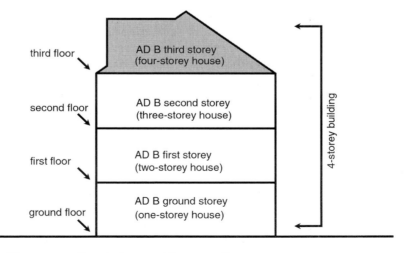

Fig. 4.4 Storey conventions in Approved Document B.

FIRE SAFETY: COMMON CONFIGURATIONS FOR SINGLE FAMILY DWELLINGHOUSES

New floor *less than* 4.5 m above lowest ground level

Generally, this is a one-storey dwelling (i.e. a bungalow) made into a two-storey dwelling. Approved Document B (AD B) *Fire safety* offers relatively few recommendations specific to loft conversions in buildings of this sort. However, it should be noted that AD B states that the guidance for a typical one- or two-storey dwelling is 'limited to the provision of smoke alarms and to the provision of openable windows for emergency egress' (AD B 1.i).

As noted earlier, the primary escape dependency in terms of safety is height. In most cases, the new floor of a bungalow loft conversion would not be more than 4.5 m above ground level. There is a presumption that self-rescue is possible at this relatively low level and the Approved Document notes that the means of escape in a typical dwellinghouse of this sort are relatively simple to provide (AD B 2.1). The guidance is always more onerous where a floor is *more* than 4.5 m above ground level.

Escape windows

- Except for kitchens, all habitable rooms in the upper storey(s) of a house served by only one stair should be provided with a suitable window or external door (AD B 2.7).

Specification of suitable windows and doors is described under 'emergency egress' in *Elements and definitions* (p. 68). Note that any window of appropriate dimensions and position may be an emergency egress window, not just a roof window (Fig 4.5a).

- A single egress window can be accepted to serve two rooms provided that *both* rooms have their own access to the stairs. A communicating door between the rooms must be provided so that it is possible to gain access to the window without passing through the stair enclosure (AD B 2.7).

Where an emergency egress window is to be a roof window, the principle of self-rescue rather than ladder-assisted rescue should be held in mind. Note, however, that no maximum window-to-eaves distance is indicated in the guidance for a floor less than 4.5 m above ground level.

- A window or door should enable a person escaping to reach a place free from danger from fire. This is a matter of judgement in each case, but in general a courtyard or back garden from which there is no exit other than through other buildings would have to be at least as deep as the dwelling is high to be acceptable (AD B 2.11b).

Because self-rescue rather than ladder-assisted rescue is anticipated in a two-storey dwelling, the thrust of the guidance is that an emergency egress window

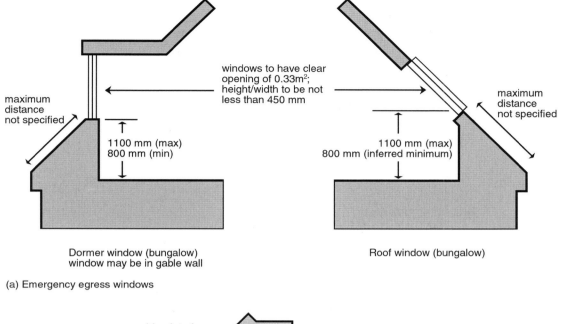

maximum distance not specified

windows to have clear opening of 0.33m²; height/width to be not less than 450 mm

maximum distance not specified

1100 mm (max)
800 mm (min)

1100 mm (max)
800 mm (inferred minimum)

Dormer window (bungalow)
window may be in gable wall

Roof window (bungalow)

(a) Emergency egress windows

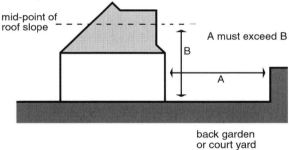

mid-point of roof slope

A must exceed B

B

A

back garden or court yard

(b) Escape to enclosed space

Fig. 4.5   Bungalow loft conversion.

need not be accessible from the highway. Given that this is the case, the relationship between the depth of any enclosed space and the height of the building should be considered (Fig. 4.5b). Note that where a floor is *more* than 4.5 m above ground level, pedestrian access is a requirement below the egress window.

Stair enclosure

Under current guidance, it would generally not be necessary to provide a protected stairway when a one-storey building such as a bungalow is converted to create a two-storey dwelling, provided that (a) the new floor is less than 4.5 m above ground level and (b) emergency egress windows are provided, which they must be.

Fire resistance of new floor

Guidance specific to the fire resistance of the new floor in a bungalow loft conversion is not provided in AD B. It would therefore be reasonable to adopt at least the *modified 30-minute* standard of resistance specified for the upper storey of a two-storey house (AD B Appendix A, Table A1).

The new floor will also need to conform to guidance on sound resistance set out in Approved Document E, *Resistance to the passage of sound*. Note that the specification for achieving 40 dB airborne sound insulation (set out in AD E 5.23) might, with minor modifications, provide a somewhat higher level of fire resistance than the modified 30-minute standard recommended to satisfy fire safety requirements (see Fig. 7.18 for construction details).

Means of warning

- A mains-supplied automatic smoke detection and alarm system based on linked smoke alarms should be installed (AD B 1.8).

The system should be installed throughout the dwelling with at least one alarm for each storey. Alarms must be sited in the circulation space within 7.5 m of habitable rooms and more than 300 mm from any walls or light fittings.

New floor *more than* 4.5 m above ground level

Generally, this is a two-storey dwelling made into a three-storey dwelling. Currently, the Approved Document provides three sets of guidance that are relevant when considering escape provisions for a loft conversion that forms an additional storey in an existing two-storey dwellinghouse, i.e. making it a three-storey dwelling with a new floor that is more than 4.5 m above ground level. For the sake of clarity, these procedures are described below as *enclosed stairway measures*, *protected stairway measures* and *alternative escape route measures* (note that the Approved Document does not use these exact expressions).

- *Enclosed stairway measures* are intended *specifically* for use with loft conversions under a closely defined but relatively common set of circumstances. This guidance is unique in that it permits reduced protection to the stairway and a presumption of ladder rather than self rescue. Note that this approach may be withdrawn during 2006.
- *Protected stairway measures* represent the conventional approach to fire safety and provide the highest standard of passive safety.
- *Alternative escape route measures* remove the need for a fully protected stairway, provided the top storey is separated by fire-resisting construction and has an alternative escape route via a final exit (but not a window).

In all three cases, the conditions are considerably *less* onerous if the house has more than one internal stairway which affords an effective alternative means of escape from the loft, provided the stairways are adequately separated from each other.

Enclosed stairway measures

This approach is unique in that it applies only to loft conversions where a two-storey dwelling becomes a three-storey dwelling (Fig. 4.6). Note that this approach may be withdrawn during 2006. The enclosed stairway method is widely used because its provisions cover the most common sorts of loft conversion. No height above lowest ground level is specified, although it would be reasonable to assume that the new floor would be more than 4.5 m and less than 7.5 m above lowest ground level. This method is *not* applicable if the new storey has a floor area that exceeds 50 m² or if the new storey is to contain more than two habitable rooms (AD B 2.17a/b).

Enclosed stairway measures generally allow existing doors to the stair enclosure to be retained, provided that they are made self-closing. It is based on the provision of fire-resisting construction to separate the new floor from the rest of the house; in a fire, this allows the occupants of the new floor to wait in relative safety for rescue via a ladder through a suitably placed window. It is emphasised that a window cannot normally be considered as a primary means of escape for floors more than 4.5 m above ground level, but this guidance represents an exception.

The enclosed stairway 'standard' may not be used in new-build situations and only applies where the conversion creates a three-storey dwelling. It generally does not apply where a roof is substantially replaced. See also 'loft conversion' in *Elements and definitions*, p. 68.

Enclosure of the existing stair

■ The stair in the ground and first storeys should be enclosed with walls and/or partitions which are fire resisting, and the enclosure should either: (a) extend to a final exit or (b) give access to at least two escape routes at ground level, each delivering to final exits and separated from each other by fire resisting construction and self-closing fire doors* (AD B 2.18).

Figures 4.7a and 4.7b indicate acceptable relationships between the stair and final exits, while Fig. 4.7c illustrates a commonly encountered but unacceptable 'open-plan' relationship that would generally mean a loft conversion could not be carried out.

Stairways that pass through and over habitable rooms, which are common in dwellings with a central stair, must be carefully considered (Fig. 4.9b). Soffits and any other elements enclosing the stair must provide 30-minute fire resistance. Because existing doors of unknown fire resistance are retained under 'enclosed stairway' measures, it does not provide the same standard as a protected stairway. See also 'final exit' in *Elements and definitions* (below).

---

* By virtue of AD B 2.19 and the note to Diagram 3 of AD B, however, there is a presumption that existing doors need only be made self-closing under the 'enclosed stairway' approach.

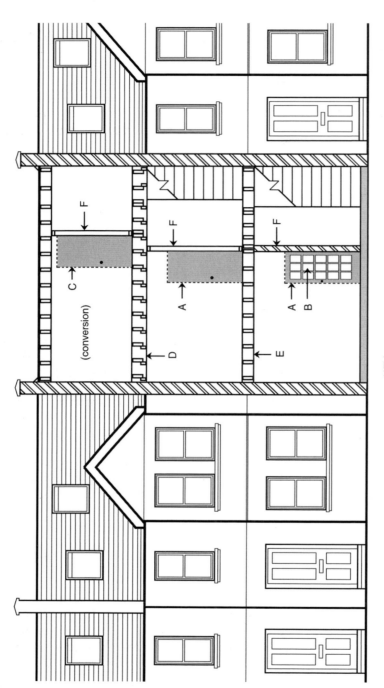

KEY:

A - existing doors made self closing

B - glazing to enclosure to be fire resisting

C - any new doors to FD20 standard;
    fire separation door to conversion must be FD20

D - new second floor: 30-minute fire resistance

E - existing first floor: modified 30-minute fire resistance

F - stair enclosure: 30-minute fire resistance

Fig. 4.6 'Enclosed stairway' method: fire resistance of enclosure.

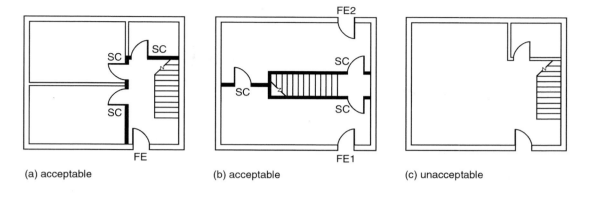

(a) acceptable   (b) acceptable   (c) unacceptable

KEY:
SC - existing door made self closing
FE - final exit

■■■■■ - indicates 30-minute
fire-resisting construction

Fig. 4.7 'Enclosed stairway' method: stair enclosure and escape routes at ground level.

Doorways

■ Every doorway within the enclosure to the existing stair should have a door which, in the case of doors to habitable rooms, should be fitted with a self-closing device. Rising-butt hinges are adequate as self-closing devices (AD B 2.19).

See also 'automatic self-closing device' in *Elements and definitions* (p. 68).

Doors

■ Any *new* door to a habitable room should be an FD 20 fire door (but see 'fire doors' in *Elements and definitions* (p. 68). Existing doors need only be fitted with self-closing devices. Existing glazed doors may need to have the glazing changed (AD B 2.19).

This concession applies only to loft conversions under 'enclosed stairway' measures.

Glazing

■ Any glazing (whether new or existing) in the enclosure to the existing stair, including all doors (whether or not they need to be fire doors), but excluding glazing to a bathroom or WC, should be fire-resisting and retained by a suitable glazing system and beads compatible with the type of glass (AD B 2.20).

See also 'glazing' in *Elements and definitions* (p. 68). Note that it may be prudent to consider the fire resistance of a conventionally glazed door or fanlight for a bathroom containing a boiler.

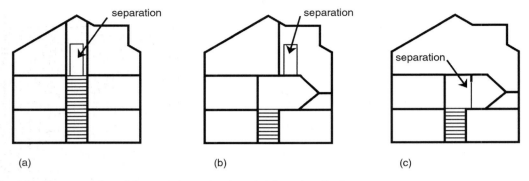

Fig. 4.8 Fire separation of the new storey: 'enclosed stairway' method.

New stair

- The new storey should be served by a stair (which may be an alternating tread stair or fixed ladder) meeting the provisions of Approved Document K *Protection from falling, collision and impact*. The new stair may be located either in a continuation of the existing stairway, or in an enclosure that is separated from the existing stairway, and from ground and first floor accommodation, but which opens in the existing stairway at first floor level (AD B 2.21). See also Fig. 4.8.

Fire separation of new storey

- The new storey should be separated from the rest of the house by fire-resisting construction. To maintain this separation, measures should be taken to prevent smoke and fire in the stairway from entering the new storey. This may be achieved by providing a self-closing fire door set in fire-resisting construction at either the top or the bottom of the new stair, depending on the layout of the new stairway (AD B 2.22). See also Fig. 4.8.

The purpose of providing separation is to allow the occupants of the conversion to await rescue in relative safety.

First floor fire resistance

- Under 'enclosed stairway' measures, the existing *first* floor need not be upgraded to the full 30-minute fire resistance standard, provided that (a) only one storey is being added, (b) the new storey contains no more than two habitable rooms and (c) the total area of the new storey does not exceed 50 m². Under these circumstances, the modified 30-minute standard is acceptable for the first floor, provided that it separates only rooms (and not circulation spaces) and that all the conditions of 'enclosed stairway' measures set out here are adhered to.
- The floor needs to meet the full 30-minute standard where it forms part of the enclosures to the circulation space between the loft conversion and the final exit (AD B 8.7). See also Fig. 4.9a.

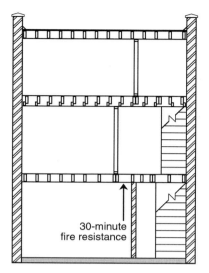

Fig. 4.9a 'Enclosed stairway' method: floor forming part of enclosure to the circulation space.

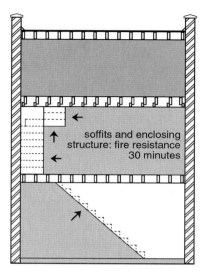

Fig. 4.9b Stair passing through or over habitable rooms (all conversions).

Note that under normal circumstances, *all* floors in a three-storey building are required to meet the full 30-minute standard of fire resistance. The acceptance of the modified 30-minute standard under certain circumstances thus represents an important concession. See also 'modified 30-minute' in *Elements and definitions*, p. 68.

Escape windows

■ The room (or rooms) in the new storey should each have an openable window or roof light which meets the relevant provisions (see Fig. 4.10). A door to a roof terrace is also acceptable.

■ In a two-room loft conversion, a single window can be accepted provided that both rooms have their own access to the stairs. A communicating door between the rooms must be provided so that it is possible to gain access to the window without passing through the stair enclosure (AD B 2.24).

■ The window should be located to allow access for rescue by ladder from the ground. There should therefore be suitable pedestrian access to the point at which the ladder would be set, for fire service personnel to carry a ladder from their vehicle, although it should not be assumed that only the fire service will make a rescue.

■ Escape across the roof of a ground storey extension is sometimes acceptable providing the roof is fire-resisting (AD B 2.5). The effect of an extension on the ability to escape from windows in other parts of the house (especially from a loft conversion) should be considered (AD B 2.25).

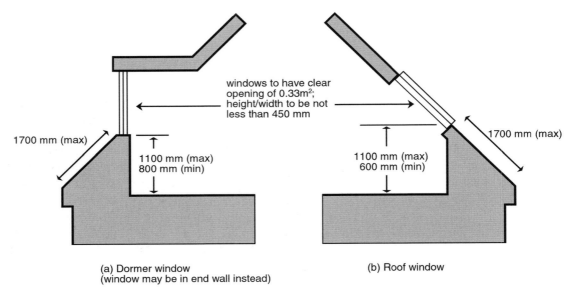

windows to have clear
opening of 0.33m²;
height/width to be not
less than 450 mm

1700 mm (max)

1100 mm (max)
800 mm (min)

1100 mm (max)
600 mm (min)

1700 mm (max)

(a) Dormer window
(window may be in end wall instead)

(b) Roof window

Fig. 4.10   'Enclosed stairway' method: emergency egress windows.

Self-rescue is seldom practical where floor heights exceed 4.5 m above lowest ground level and, normally, emergency egress windows would not be provided at this height. Under 'enclosed stairway' measures, however, emergency egress window(s) are positioned to facilitate ladder rescue. Note that a fixed ladder on the slope of the roof is not recommended. 'Enclosed stairway' measures represent the *only* set of circumstances in the guidance where an emergency egress window is acceptable for high-level rescue.

Particular care should be taken if any alteration to the floor plan is made during the conversion. Changes to room layouts made during construction sometimes lead to an emergency egress window being left *outside* the conversion's fire separation envelope (for example, on a second floor landing on the escape route). Such an arrangement would not be acceptable.

Automatic smoke detection and alarms

- A mains-supplied automatic smoke detection and alarm system based on linked smoke alarms must be installed throughout the building (AD B 1.8).

At least one alarm should be provided for each storey. The arrangement of the system is dependent on the layout of the dwelling. Alarms must be sited in the circulation space within 7.5 m of habitable rooms and more than 300 mm from any walls or light fittings.

Air circulation systems

See *Elements and definitions*, p. 68.

Passenger lifts

See *Elements and definitions*, p. 68.

Protected stairway measures

This method offers the highest standard of passive fire protection for houses with a floor more than 4.5 m above ground level, and is the same standard that is applied to new dwellings of three storeys (Fig. 4.11). It must be used where the new storey exceeds 50 m² in floor area or where the new storey will contain more than two habitable rooms. It would also be reasonable to apply this standard where the new storey substantially replaces an existing roof structure (see *Elements and definitions*, p. 68, for a definition of loft conversion).

With this approach, a protected escape route formed with fire-resisting construction and fire-resisting, self-closing doors leads down through the house to the final exit. This approach does not depend on the provision of emergency egress windows in the roof.

Protected stairway

■ The upper storeys (those above the ground storey) should be served by a protected stairway which should either: (a) extend to a final exit or (b) give access to at least two escape routes at ground level, each delivering to final exits and separated from each other by fire-resisting construction and self-closing fire doors (AD B 2.13a). See also Fig. 4.12.

Stairways that pass through and over habitable rooms, which are common in dwellings with a central stair, must be carefully considered (Fig. 4.9b). Soffits and any other elements enclosing the stair must provide 30-minute fire resistance.

Doors

■ Any door forming part of the enclosures to a protected stairway in a single-family dwellinghouse should have a minimum fire resistance to FD 20 standard (AD B, Appendix B Table B1). See also 'fire doors' in *Elements and definitions*, p. 68.

Doorways

■ All fire doors forming part of the enclosures to a protected stairway should be fitted with an automatic self-closing device. Rising-butt hinges are adequate in this application.

Glazing

■ All glazing within the protected stairway enclosure, including that in doors, walls and fanlights, to be fire-resisting.

See also 'glazing' in *Elements and definitions*, p. 68.

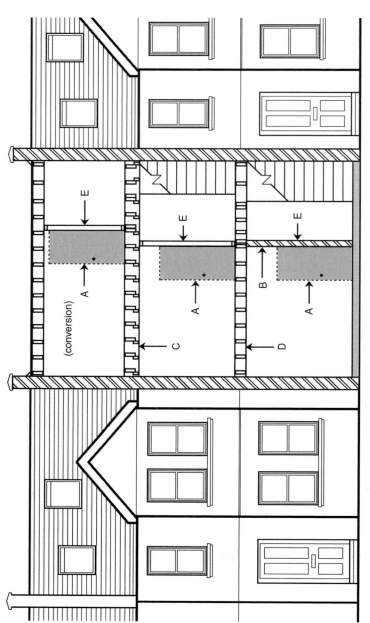

KEY:
A - doors to be FD20 self-closing
B - glazing to enclosure (e.g. doors, fanlights) to be fire resisting
C - new second floor: 30-minute fire resistance
D - existing first floor: 30-minute fire resistance
E - stair enclosure: 30-minute fire resistance

Fig. 4.11   'Protected stairway' method: enclosure and floor fire resistance.

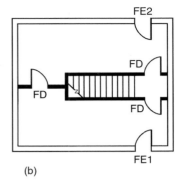

(a)

(b)

KEY:
FD - FD20 self-closing fire door
FE - final exit

■■■■■ - indicates 30-minute
fire-resisting construction

Fig. 4.12 'Protected stairway' method: stair enclosure and escape routes at ground level.

New stair

- New storey to be served by a stair meeting the provisions of Approved Document K *Protection from falling, collision and impact*. The new stair should be located within the protected stairway enclosure.

Fire separation of new storey

This does not generally apply in the sense described in the 'enclosed stairway' method. The approach adopted in this section provides a higher standard of passive fire protection and, because this means that the occupants of the conversion need not be provided with a 'safe haven' to await ladder rescue, loft-specific fire separation, as described earlier, is not provided.

First floor fire resistance

- Floors both old and new must provide the full 30-minute standard of fire resistance (AD B 8.7 and AD B Appendix A Table A2). See also Fig. 4.11.

The modified 30-minute standard that is accepted under the circumstances set out in 'enclosed stairway' measures (above) is a concession and it does not apply under this method.

Escape windows

These are not generally required. The safety of the occupants in the conversion is not dependent on emergency egress windows where a protected stairway is provided.

Automatic smoke detection and alarms

- A mains-supplied automatic smoke detection and alarm system based on linked smoke alarms must be installed throughout the building (AD B 1.8).

At least one alarm should be provided for each storey. The arrangement of the system is dependent on the layout of the dwelling. Alarms must be sited in the circulation space within 7.5 m of habitable rooms and more than 300 mm from any walls or light fittings.

Air circulation systems

See *Elements and definitions*, p. 68.

Passenger lifts

See *Elements and definitions*, p. 68.

Alternative escape route measures

Another possible approach is to separate the conversion from the lower storeys with fire-resisting construction and provide it with an alternative escape route, such as a separate additional stairway leading to its own final exit (Fig. 4.13). The

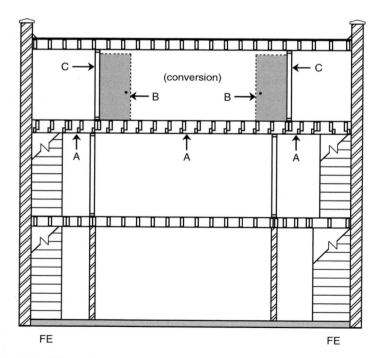

KEY:
A - new second floor requires 30-minute fire resistance

B - FD20 doors to each escape route

C - conversion separated from lower storeys by 30-minute fire-resisting construction

FE - final exit

Fig. 4.13   'Alternative escape route' method: fire separation.

alternative stairway could be internal or external. This approach is outlined in AD B 2.13b. As with 'protected stairway' (above), this method is also potentially appropriate where the new floor exceeds 50 m² or has more than two habitable rooms.

This approach is similar to 'enclosed stairway' measures set out at the beginning of this chapter, but only in as much as the conversion is separated from the storeys below by fire-resisting construction. The critical difference is that under 'alternative escape' measures, the alternative escape route must lead to a final exit rather than an emergency egress window (which is not acceptable as a final exit). Therefore even if there were to be an emergency egress window on a lower floor (with a floor less than 4.5 m above ground level) and even if it were accessible by way of an alternative independent stair, it could still not be considered a final exit.

Note that Appendix E of AD B provides a definition of 'alternative escape route' (see *Elements and definitions*, p. 68) which includes a balcony or flat roof used to reach a place of safety. The relationship between a balcony or flat roof and the place of safety would, of course, have to be carefully considered.

A suitable mains-supplied automatic smoke detection and alarm system based on linked smoke alarms must be installed throughout the building (AD B 1.8).

Given that most single-family dwellinghouses have only (and have only space for) a single stair and, given that external fire escapes are rarely permitted under planning legislation, the alternative escape route approach is seldom adopted.

New floor at or more than 7.5 m above ground level

Generally, this is where a three-storey dwelling is made into a four-storey dwelling. It is potentially the most complex form of conversion as far as providing a means of escape is concerned (Fig. 4.14). Approved Document B states the following:

'Houses with more than one floor over 4.5 m above ground level
Where a house has two or more storeys with floors more than 4.5 m above ground level (typically a house with four or more storeys), then in addition to meeting the provisions in paragraph 2.13 [*protected stairway or separation and alternative escape route for top floor*], an alternative escape route should be provided for each storey or level situated 7.5 m or more above ground level.

Where the access to the alternative escape route is via:
(a) the protected stairway to an upper storey; or
(b) a landing within the protected stairway enclosure to an alternative escape route on the same storey;
then the protected stairway at or about 7.5 m above ground level should be separated from the lower storeys or levels by fire-resisting construction' (AD B 2.14).

This approach is based on the provision of an alternative escape route. An escape route is defined as a 'route forming that part of the means of escape from any point in the building to a final exit' (AD B Appendix E). Strictly interpreted, this rules out the use of emergency egress windows because the definition of 'final exit' in the same Appendix notes that windows are not acceptable as final exits.

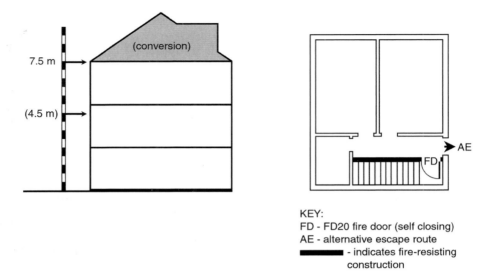

Fig. 4.14   Floor at 7.5 m above ground level.

From a theoretical perspective, therefore, the only way to meet the current guidance (both in AD B and in BS 5588 Part 1) is to provide an additional independent stairway leading to a place of safety. This would have to be provided either inside the building (where it must be separated from the main stair by fire-resisting construction) or outside the building (an external fire escape).

Whilst this would represent the most satisfactory approach, it is often not feasible. As stated earlier, external fire escapes are generally not acceptable for planning reasons, and most dwellings are not capable of accommodating an additional internal stairway. Given that the Approved Documents are guidance rather than edict, it is perhaps not surprising that various forms of alternative semi-official guidance have emerged to satisfy Requirement B1.

Although it is not the purpose of this book to advocate the use of such guidance, the attention of readers is drawn to the fact that many local authorities are prepared to consider alternative approaches (trade-offs) where it is not possible to comply with the guidance in AD B or BS 5588. Note that a 'trade-off' is *not* the same as a relaxation: a trade-off proposal must provide the same standard of safety as that set out in the Approved Document in order to satisfy Requirement B1 ('an appropriate means of escape').

It is emphasised that 'trade-off' proposals must first be approved by a local authority under a Full Plans application before work commences.

The following 'trade off' proposals for a high-level loft conversion (with a floor at or more than 7.5 m above ground level) are typical of the sort that would be *considered* by a local authority for a dwelling with all its storeys less than 200 m² (Fig. 4.15). They were made by a local authority where conversions of this sort are common:

- Every door off the stairs leading to accommodation or storage to have 30-minute fire resistance and be provided with self-closure device. Bathroom

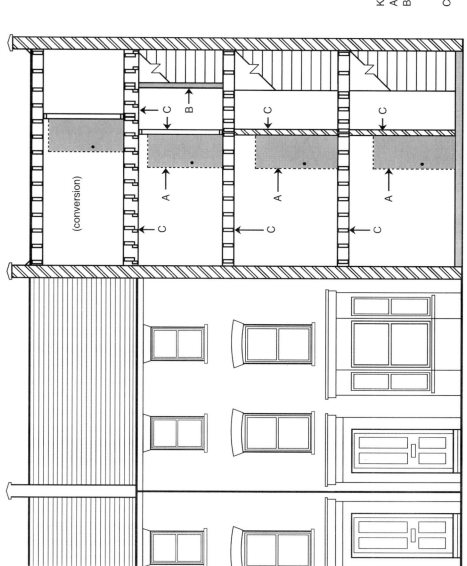

Fig. 4.15   Floor at 7.5 m above ground level: trade-off proposals.

fire resistance provided that there is no boiler in the bath-
walls separating the accommodation from the bathroom are

nclosing walls at all levels, without any glazing, with at least
:sistance to the stairs and entrance corridor leading to the front
)rs with self-closure devices).
e a 30-minute fire-resisting door and screen at second floor level
) floor stairs from below (FD 30S with self-closure device).
)n and alarm system should be provided to cover the escape route
r down to the street exit. This includes the provision of:
: detector in all rooms off the entrance hall and the stairs in every
,except bathrooms and WC room)

(ii)   heat detector in the kitchen and if boiler is in a bathroom (a smoke detector
       might give false alarms)
(iii) smoke detectors in the hall and floor landing areas of each storey
(iv)   normal cabling being used to interconnect all the detectors (fire resistant
       cabling better)
(v)   loud alarms at the top of the house to wake people asleep in their beds (75 dB
       at the head of bed – built-in sounder to smoke detector may be adequate)
(vi)   control panel with mains connection and rechargeable battery backup in
       the entrance hall.

The attention of readers is also drawn to a determination by the Secretary
of State in which a comparable approach to loft conversion fire safety is considered in a building of similar height. This is reproduced in Appendix 2 (item 9,
reference 45/1/200).

Sleeping galleries

Sleeping galleries are sometimes provided in single-storey buildings, generally to
exploit a roof void that is open to the dwelling (Fig. 4.16). It is probably the oldest
form of 'loft' conversion. Guidance on means of escape from a sleeping gallery is
provided in AD B 2.9 and it is reproduced here:
Where a sleeping gallery is provided:

- The gallery should be no more than 4.5 m above ground level;
- The distance between the foot of the access stair to the gallery and the door to
  the room containing the gallery should not exceed 3 m;
- An alternative exit, or an emergency egress window which complies with paragraph 2.11* is needed if the distance from the head of the access stair to any
  point in the gallery exceeds 7.5 m; and
- Any cooking facilities within a room containing a gallery should either:
  (i) be enclosed with fire-resisting construction; or
  (ii) be remote from the stair to the gallery and positioned such that they do not
       prejudice the escape from the gallery.

See Figs 4.5a and 4.5b.

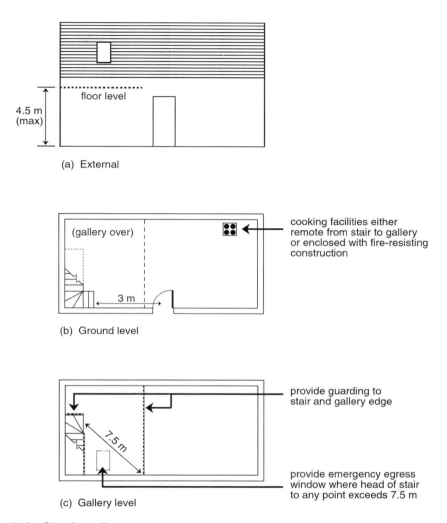

Fig. 4.16 Sleeping gallery.

Sleeping galleries have been the subject of determinations by the Secretary of State. See determination letter 45/1/213 in Appendix 2 (item 12).

Ongoing requirements

Currently, neither the Building Act 1984 nor the Building Regulations impose any meaningful obligation on householders to maintain safety systems or features. It is recognised, for example, that some householders deactivate self-closing devices and remove fire-resisting doors; smoke detection systems, once installed, may not be subject to continuing testing or adequate maintenance.

A high number of deaths and injuries occur in buildings where smoke detection and warning systems are not maintained. For example, in the London Fire Brigade area between 2002 and 2004, fitted alarm systems failed to operate in 273 recorded fire incidents. Reasons for non-actuation included disconnection and, in some cases, the removal of hard-wired systems. These fires killed 36 people and injured more than 400 (source: London Fire and Emergency Planning Authority).

ELEMENTS AND DEFINITIONS

Air circulation systems

These include air circulation systems for heating, energy conservation and condensation control in houses with a floor *more than* 4.5 m above ground level. With systems of this type, the following precautions are needed to avoid the possibility of smoke or fire spreading into a protected stairway (AD B 2.15):

(a) Transfer grilles should not be fitted in any wall, door, floor or ceiling enclosing a protected stairway.
(b) All ductwork passing through the enclosure to a protected stairway should be so fitted that all joints between the ductwork and the enclosure are fire-stopped.
(c) Where ductwork is used to convey air into a protected stairway through the enclosure of the protected stairway, the return air from the protected stairway should be ducted back to the plant.
(d) Air and return air grilles or registers should be positioned at a height not exceeding 450 mm above floor level.
(e) A room thermostat for a ducted warm air heating system should be mounted in the living room at a height between 1370 mm and 1830 mm and its maximum setting should not exceed 27°C.
(f) Any system of mechanical ventilation which recirculates air should comply with the relevant recommendations in BS 5588 *Fire precautions in the design, construction and use of buildings*, Part 9 *Code of practice for ventilation and air conditioning ductwork*.

Alternative escape route

Escape routes sufficiently separated either by direction and space, or by fire-resisting construction, to ensure that one is still available should the other be affected by fire. Note: a second stair, balcony or flat roof which enables a person to reach a place free from danger from fire is considered an alternative escape route for the purposes of a dwelling house (AD B Appendix E). See also 'balconies and flat roofs' (p. 69).

Automatic self-closing devices (self closers)

Appendix E to Approved Document B defines automatic self-closing devices as being capable of closing the door from any angle and against any latch fitted to the door. However, the guidance notes that rising-butt hinges which do not meet those criteria are acceptable for doors to or within dwellings. Fire doors and existing doors in stairways may be fitted with either spring or rising-butt self-closing devices.

Balconies and flat roofs

A flat roof or balcony may form part of a means of escape. To do so, it must comply with the following three provisions:

- The roof should be part of the same building from which the escape is being made
- The route across the roof should lead to a storey exit or external escape route
- The part of the roof forming the escape route and its supporting structure, together with any opening within 3 m of the escape route, should provide 30-minute fire resistance

Guarding may be needed where a balcony or flat roof is provided for escape purposes. Guidance provided in Approved Document K *Protection from falling, collision and impact* indicates that a guarding height of 1100 mm is appropriate for external balconies and roof edges in single family dwellings.

Doors – glazing in final exit

In general, it is not necessary to replace a final exit door (such as a front door facing a highway) with one of known fire resistance. However, in cases where a final exit might constitute an unprotected area (for example, a door in the flank wall of a building), it should be considered relative to guidance provided in section 14 of Approved Document B.

Emergency egress windows

Emergency egress windows, sometimes described as means of escape or MoE windows, should have an unobstructed openable area that is at least 0.33 m^2. Neither the openable width nor height should be less than 450 mm (note that, to achieve 0.33 m^2 with a width of 450 mm, the window height would have to be at least 740 mm). The major manufacturers of roof windows generally indicate which of their products are suitable in emergency egress applications. It should be noted that *any* appropriately positioned window is capable of serving as an emergency egress window (a dormer window, for example) provided it has an openable area of the appropriate size.

Approved Document B makes a distinction between emergency egress windows provided for self-rescue (for example, in basement, ground floor and first floor situations) and emergency egress windows in a loft conversion that turns a two-storey dwelling into a three-storey dwelling. Safe use of the emergency egress window in the latter case depends on a ladder being set up.

- For floor levels up to 4.5 m above ground level see Fig. 4.5a
- For 'enclosed stairway' loft conversions see Fig. 4.10. Note the maximum window-to-eaves distance criterion and minimum height without the need for guarding.

In all cases, the bottom of the window's openable area should not be more than 1100 mm above the floor. Where it is not possible to fix the window less than

1100 mm above floor level (perhaps because of an obstructing purlin), it is sometimes possible to provide a permanent fixed step. However, in the case of a loft conversion of the sort described in the 'enclosed stairway' method, this would not remove the requirement for the window-to-eaves distance to be 1700 mm or less. (References in AD B: 2.11 and 2.23–2.25).

Escape route

Route forming that part of the means of escape from any point in a building to a final exit (AD B Appendix E).

Final exit

The termination of an escape route from a building giving direct access to a street, passageway, walkway or open space, and sited to ensure rapid dispersal of persons from the vicinity of a building so that they are no longer in danger from fire and/or smoke. Note: windows, including emergency egress windows, are not acceptable as final exits (AD B Appendix E).

In the case of domestic loft conversions, the final exit is generally the front door, or the front door *and* back door in situations where two exits are required.

Fire doors

When considering the fire resistance of doors, the resistance of the *entire* assembly must be taken into account including frame and furniture. It is emphasised that, in all cases, the fire resistance of the door assembly is dependent upon the use of intumescent strips in the door or frame. Some manufacturers produce doors with integral intumescent strips, but not all. Common fire door specifications for loft conversions outlined in this chapter (tested to BS 476 part 22) include:

- **FD 20:** Generally specified for use in forming part of the enclosures to a protected stairway in a single-family dwellinghouse. Note that although the FD 20 standard still exists, relatively few FD 20 doors are now manufactured, and doors of FD 30 standard are used instead.
- **FD 30:** Specified for use between a dwellinghouse and a garage. Now widely used in place of FD 20 doors (see above).
- **FD 30S:** Smoke control fire doors may be required in certain circumstances (see *New floor at or more than 7.5 m* for an example).

Glazing on escape routes

Because not all fire-resisting glazing is designed to reduce the transfer of heat, Approved Document B sets limits on the use of uninsulated glazed elements on escape routes (although the glazing must, of course, be fire-resisting). Table A4 of AD B sets out these limitations in detail. Note that, when considering fire-resisting glazing, the fire resistance of beading must also be taken into account.

Habitable room

A room used, or intended to be used, for dwelling purposes. For the purposes of Approved Document B *Fire safety*, a habitable room includes a kitchen but not a bathroom (AD B Appendix E).

Loft conversion – definition for the purposes of 'enclosed stairway' measures

Approved Document B does not provide a definition of loft conversion. While paragraph 2.17 refers to the conversion of the *existing* roof space into habitable rooms, diagram 6 in the same document illustrates a dormer window projection. As a matter of general interpretation, therefore, a rear box dormer is generally considered to be a loft conversion for the purposes of 'enclosed stairway' procedures set out earlier in this chapter (provided the floor area and room number rules are observed).

However, work that substantially *replaces* an existing roof structure – a new front and rear mansard for example – is often not considered to constitute a loft conversion. Although it refers to a specific case, attention is drawn to the following extract from an appeal dated 26 July 2001 (reference 45/3/148): 'The guidance in Approved Document B for loft conversions relates only to the conversion of an existing roof space and, as such, would not be applicable to this case as the new habitable rooms are being created by a vertical extension and the replacement of the existing roof by a mansard roof structure.'

Modified 30-minute fire resistance

The modified 30-minute standard for floors satisfies the test criteria for the full 30 minutes in respect of load-bearing capacity, but allows reduced performances for integrity and insulation. Generally, it applies only to first-floor construction in two-storey dwellings. However, it is also permitted where a loft conversion creates a three-storey dwelling under conditions set out in the 'enclosed stairway' measures. As noted in Chapter 7, the term 'modified' does not necessarily mean that anything need be done to an existing floor, provided that it is capable of meeting the standard. Note that the floor for the conversion itself (the new second floor) would require full 30-minute fire resistance.

Open plan layouts

Open plan layouts, often adopted at ground floor level in single-family dwellings, have grown in popularity since the early 1970s. These fall broadly into two types:

■ *Through lounge*. In a typical 'through lounge' or 'knock-through' configuration, the wall separating a pair of previously independent habitable rooms, perhaps a kitchen and a dining room, is removed to create a large single space. Provided that the new, larger, room is *not* open to the stairway, and doorways and doors have been retained, such an arrangement would probably not prejudice the conversion of the roof space.

■ *Open plan*. In a full 'open plan' configuration (Fig. 4.7c), however, the walls separating habitable rooms from the stair are not present. It should be noted that fires tend to originate in habitable rooms rather than circulation areas and, where habitable rooms and circulation areas have been merged by the removal of partition walls, the risks to the occupants of the building from fire and smoke are considerably increased. A loft conversion in such a building is therefore generally not possible.

Passenger lifts

Where a passenger lift is provided in the house and it serves any floor more than 4.5 m above ground level, it should either be located in the enclosure of the *protected stairway* (see below) or be contained in a fire-resisting lift shaft (AD B 2.16).

Protected stairway

A stair discharging through a final exit to a place of safety, including any exit passageway between the foot of the stair and the final exit, that is adequately enclosed with fire resisting construction. References to protected stairway are found in AD B 2.13 and AD B Appendix E.

There is an important difference between a *protected stairway* in which the fire-resisting construction includes fire doors, and the *enclosure of an existing stair* allowed in certain circumstances (described earlier in this chapter) when a loft conversion is carried out. When an existing stair is enclosed, existing doors need only be fitted with self-closing devices. Only new doors must be fire doors. An 'enclosed' stairway of this sort, although offering a degree of fire resistance, should not be considered a protected stairway.

Residential sprinkler systems

Automatic sprinkler systems for residential use are becoming more widely used in the UK. Guidance published by some local authorities indicates that there is an increased willingness to consider a lower standard of passive fire protection if this is traded off against active fire protection, such as a whole-house sprinkler system fitted as part of a loft conversion.

Where a residential sprinkler system is installed it *might* not be necessary for new doors to be fire doors, any doors to be fitted with self-closing devices or to upgrade the fire resistance of the walls or floors. An alarm system would still be required and, where appropriate, escape windows provided.

Where such guidance is published, it represents general advice only and is the view of a particular local authority. It is not necessarily universally accepted. Sprinkler systems for domestic use are designed and installed in accordance with BS 9251:2005 *Sprinkler systems for residential and domestic occupancies. Code of practice.*

Self-rescue

There is a presumption that self-rescue, that is to say, rescue without the intervention of a third party, is possible from a room where the floor is less than 4.5 m above ground level.

Storey exit

A final exit, or a doorway giving direct access into a protected stairway or external escape route (AD B Appendix E).

Storey height measurement

Means of escape provisions are based on a number of factors including the height of the top floor relative to ground level. Ground level is measured at the lowest side of the building (Fig. 4.3).

Subsequent alterations to dwellings

Any proposal to create an open-plan configuration for lower storeys *after* a loft conversion has been carried out is likely to be rejected.

APPROVED DOCUMENT B: PROPOSED CHANGES TO GUIDANCE

Guidance on loft conversions contained in Approved Document B is undergoing major changes with new guidance due to be launched in 2006. Proposed changes include:

- Specific loft conversion guidance ('enclosed stairway' measures) to be withdrawn because it is considered anomalous.
- Fully protected stairway with fire doors required for *all* loft conversions, including bungalows. Self-closers, though, may no longer be required. Modified 30-minute standard may be retained (two rooms or less than 50 m²).
- Sprinklers rather than alternative escape for high-level conversions (BS 9251).
- 'Variation of provisions' which explicitly allows a building control body (such as a local authority) to take into account fire safety features that are not included in the Approved Document.
- Heat alarms/detectors required for material alterations such as loft conversions.
- Smoke alarm to be provided in largest bedroom.
- Mains-powered smoke alarms may be interconnected using radio links.
- Sleeping gallery guidance to change, with the word 'sleeping' omitted and a new definition of 'gallery' provided: 'a raised area or platform around the sides or at the back of a room which provides extra space'.
- Galleries higher than 4.5 m may be possible.
- Emergency egress windows: minimum of 600 mm clarified for bungalows.
- Locks on emergency egress windows considered.
- Cavity barriers to enclosures above protected stairways.

5 Conversion survey

As noted in Chapters 3 and 4, the form of the existing building and its roof type will have a considerable influence on the design of the conversion. The condition and structural configuration are also of critical importance. At the preliminary stage of the conversion process, therefore, it is necessary to conduct a survey to identify:

- The suitability of the roof space for conversion
- The internal arrangement of the building
- The structural configuration of the building
- The condition of relevant building fabric
- The extent of any remedial work that is required and any other work that could profitably be carried out at the same time as the conversion

A loft conversion is sometimes not feasible, and a properly conducted survey will reveal this at an early stage. Where a conversion cannot be carried out, it is generally for one or more of the following reasons:

- *Open-plan layout*: where the conversion would create a floor more than 4.5 m above ground level, the stair escape route must be protected by fire resisting construction (see Chapter 4). In a building with an open plan layout, therefore, it would be necessary either to reinstate partitions or to find some other way of protecting the stair *throughout* the building.
- *Headroom*: there is no minimum height within rooms but 2.3 m is considered to be reasonable. Lack of headroom is a potentially intractable problem: raising the ridge height of a roof is not always feasible from a planning perspective. Equally, dropping ceilings in the floor below is often not economically viable. Note that a low final ceiling height (or lower than expected) is one of the commonest causes of dissatisfaction when lofts are converted.
- *Roof structure*: it is not always considered to be cost-effective to adapt a trussed rafter roof, although it is usually technically feasible provided there is adequate headroom available.

SURVEY PROCEDURE

To proceed with the design of a loft conversion, it is necessary first to produce or acquire detailed drawings for the whole of the existing building with full construction details and measurements.

The importance of examining the whole building cannot be overstated. The primary reason for this is that elements remote from the conversion itself can have an impact on feasibility. For example, the relationship between the stairway, rooms and final exits at ground floor level will have a determining influence as far as fire safety is concerned.

In some instances, especially where the building is of recent construction, original plans may be available. It should be noted, however, that there is sometimes a considerable difference between what appears on paper and what was actually built. In addition, where original plans are to be used, any subsequent alterations to layout should be carefully noted. Note that where it is necessary to produce drawings from scratch, as it is in most cases, plans, elevations and sections should be at a scale of not less than 1:50.

Where drawings are to provide the basis of structural calculations, detailed measurements should be provided. These might, for example, include elements such as the span, spacing and dimensions of rafters and purlins. Comprehensive and accurate measurements provided on drawings may allow a structural engineer to perform some calculations without the need for a site visit.

The survey must take into account factors that may affect the viability of the conversion and this includes the general condition of relevant building elements. It should also identify any additional works that could profitably be carried out at the same time as the conversion, for example, retiling a front roof slope or repointing masonry (Fig. 5.1).

Fig. 5.1 Re-pointing stacks. Conversion provides an opportunity to carry out remedial work.

OUTLINE OF MAJOR ELEMENTS OF A SURVEY

The major survey elements are listed below. A more detailed examination of these is provided in the following pages.

Age of building
- Date of construction (including construction dates of subsequent extensions)

Headroom
- Available headroom in roof space
- Optimum positions for new staircase and landing
- Position and form of existing staircase(s)

External relationships
- Position and size of adjacent buildings/structures
- Ground level extensions (existing or proposed) to the dwelling
- Pedestrian/ladder access (buildings of two storeys and over)
- Loft 'terracing'

Internal layout
- Position, use and dimensions of rooms throughout the dwelling
- Positions of final exits

Roof form
- External form and plan of roof

Roof structure
- System of construction (cut roof, TDA truss or trussed rafter)
- Positions, dimensions, spacing and spans of all elements of roof structure

Roof condition
- Evidence of fungal and insect attack
- Evidence of water penetration
- Fixity of structural elements
- Condition of roofing material, fixings and underlay
- Condition of valleys, soakers and flashings

Walls
- Load-bearing walls
- Gable walls and chimneys
- Solid and cavity walls
- Lintels

Foundations
- Assessment
- Geotechnical factors
- Historical factors

Internal partition walls
- Materials and construction method
- Load-bearing internal partition walls and their foundations

- Openings (e.g. knock-throughs) and beam provision
- Internal doors and glazing to circulation areas

Floor and ceiling structure
- Ceilings (materials, construction and thickness) including floorboard gaps
- Floor and ceiling joist direction, dimensions, spacing and span
- Strength of existing timber elements

Water tanks
- Cold water storage
- Feed and expansion

Drainage and services
- Position of soil and vent pipe
- Position and adequacy of existing rainwater drainage
- Position of service pipes
- Position of ancillary elements (e.g. aerial)

Chimneys
- Chimney breast continuity between floors (internal)
- Position and dimensions of chimney breast in roof space (internal)
- Stack position and height relative to proposed conversion (external)

SURVEY ELEMENTS IN DETAIL

Age of the building

The age of the building will provide an indication of likely modes of construction (Fig. 5.2). Accurate dates of subsequent extensions should be obtained: note that the volume of any addition to the building since 1948 is deducted from permitted development allowances (see Chapter 1).

Headroom and floor-to-ceiling height

Available headroom is generally the primary determining factor in the feasibility of a loft conversion. The ability to 'stand up' beneath the apex of the existing roof is sometimes cited as a guide to the suitability of a loft for conversion. This is not a universally reliable test because the combined thickness of a new, separately supported floor and an insulated roof structure must be subtracted from the ceiling joist to ridge measurement. A new floor with 200 mm joists may exceed 250 mm in depth when boarding and a deflection gap are taken into account. A flat, warm roof structure with 170 mm joists may exceed 360 mm at the ridge end where the firring is thickest.

There is no longer a minimum ceiling height requirement *within* rooms, but for the comfort and safety of the occupants, it is generally considered desirable to create new rooms with a minimum floor-to-ceiling height of 2.3 m. However,

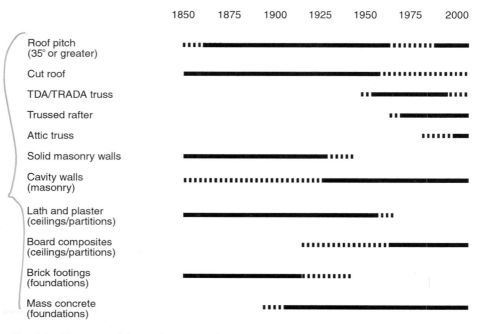

Fig. 5.2   Elements of domestic construction.

there is a presumption that any downstand elements (such as beams) within rooms present a hazard if the clearance is less than 2 m.

Approved Document K indicates explicitly that clear headroom of 2 m must be provided over the whole width of any stairway, ramp or landing. Where this cannot be achieved, the Approved Document outlines an alternative arrangement (which applies only to loft conversions) with 1.9 m to the centreline of the stairs and 1.8 m to the lowest side (Fig. 7.19b). Where the conversion provides only one habitable room, it may be possible to consider the use of an alternating-tread staircase, but only if it is not possible to accommodate a conventional stair (that is, one with a pitch not greater than 42 degrees).

In most cases, the new stair is positioned in the same stairwell as the existing stair. The new stair should enter the conversion parallel to the new floor joists to minimise the need for an extensively trimmed opening (see also Figs 7.14a and 7.21).

External relationships

The building's external relationships are important for a number of reasons. For example, an emergency egress window in the conversion of a two-storey dwelling (where the conversion creates a third storey) must be accessible by ladder from outside the building. The ground beneath such a window, therefore, must have pedestrian access and there should be adequate clearance to set up a ladder. Equally, the presence of any structures likely to impede ladder access (such as conservatories or other extensions either built or planned) should also be taken into account and indicated on drawings.

An additional consideration is the phenomenon of loft 'terracing'. A situation now often arises where a terraced house undergoing conversion may already be flanked by existing loft conversions. The survey should note the presence of adjoining conversions, and, where this is the case, a careful inspection should be made to determine the position of steel beams or other structural members.

Where steel beams have already been introduced into party/compartment walls to support an adjoining structure, it is essential to identify precisely where these have been installed. With 9" (225 mm) party/compartment walls, it is general practice to allow penetration and bearing of 100 mm, but there will clearly be many instances where penetration has exceeded this, or the wall is thinner than 9", and the potential for beam-to-beam conflict exists. The existence of disturbed brickwork at ceiling joist level may provide evidence of the position of neighbouring steelwork, but this is not always the case.

Internal layout

The relative positions of habitable rooms, the stairway and the final exit are of critical importance as far as fire safety is concerned. For the purposes of Approved Document B *Fire safety*, a kitchen is a habitable room but a bathroom is not. Note that open plan layouts are generally inimical to loft conversions (see also Chapter 4).

Roof form

Roofs are generally either gabled, with perpendicular walls built up at each end of the building, or hipped, in which case the roof slope is carried around the building. Both gable and hip roofs may incorporate projections such as bays with valleys at slope intersections. It is generally not practical to incorporate small projections such as bays within the envelope of the conversion, although these are often boarded-out for light storage. Note that, while both hip and gable roofs may be converted, it is generally simpler to convert a gabled roof.

Roof structure

The structural form of the existing roof will have a major bearing on the feasibility of the conversion. In undertaking the survey, the aim is to reach a full understanding of the structure and function of the roof. A number of examples are illustrated in Chapter 9. Note that, in some cases, the structure may defy ready analysis (Fig. 5.3).

The internal structural arrangement of a roof takes a considerable number of different forms. For the sake of convenience, roofs are often characterised as either 'cut' roofs or trussed rafter roofs. The TDA truss roof (see Fig. 9.1b) could be said to represent an evolutionary half way stage between the two.

Cut roofs often lend themselves well to conversion because, generally, a relatively small number of structural elements project into the roof void. In conducting the

Fig. 5.3   Re-conversion. This nineteenth-century roof is undergoing its second major modification in twenty years. Note the accumulation of both structural and non-structural timber elements.

survey, the position, dimensions and, where relevant, span and spacing of roof members should be recorded. These may include:

- Wall plates
- Purlins
- Purlin struts
- Straining pieces
- Ties
- Collars
- Binders
- Hangers
- Rafters (common, valley, hip, cripple, crown, jack, principal)
- Ceiling joists
- Ridge board

The position of purlins is often of particular importance because these will frequently foul the position of a means of escape window and stairway.

The direction of ceiling joists must be noted. In most cases it will be necessary to run new floor joists parallel to the existing ceiling joists, often suspended in the void between them, in order to preserve headroom.

Trussed rafter roofs are generally more troublesome to convert. With conventional trussed rafters, much of the roof void is occupied with the webs from which the roof derives a large proportion of its strength (see also Chapter 9). As well as noting truss element dimensions, method of fixing and truss spacing, the presence of bracing members should also be recorded.

Since about 2000, attic trusses (sometimes called room-in-roof or conversion trusses) have become common in new dwellings (Figs 9.1d and 9.21b). Like conventional trussed rafters, these are produced off-site by specialist manufacturers. Timber sections of considerably greater depth and breadth (for example, 47 × 197 mm rather than 35 × 97 mm sections) are used in their production. Webbing in the conventional sense is largely absent, and attic trusses therefore have the advantage of providing a useable roof void with integral floor joists.

Roof condition

A careful inspection should be made to identify any currently occurring or past damage to the roof's timber structure, including water penetration, fungal attack and insect infestation. Damaged or decayed elements may need to be replaced (Fig. 5.4). The source of any leaks, past or present, should be identified and remedied: the position of leaks is much harder to identify once insulation and final finishes have been applied. This particularly applies to front roof slopes, the coverings of which are generally not modified as part of the conversion. Evidence of fungal or insect attack will require closer scrutiny, and possibly treatment, of the entire roof structure (see also *'house longhorn beetle'*, Glossary). The condition

Fig. 5.4 Water ingress. Even a minor failure of the roof covering can cause substantial damage over time. Note that support for the purlin is provided by brick corbelling.

and adequacy of structural fixings should also be noted with particular reference to nailed connections between rafter feet and ceiling joists – these are an integral part of the tie between opposing roof slopes.

The type of roofing material should be noted, and careful attention paid to its condition. The condition of fixings and the presence or absence of underlay should also be noted. Where it is proposed to clad the conversion in a material to match the existing roof, the feasibility of reusing slates or tiles displaced by the new dormer may be considered.

The provision and adequacy of roof ventilation should also be examined. Flush eaves (see Fig. 9.3) require careful attention because they are sometimes troublesome to ventilate, and the use of proprietary tile/slate ventilators must be considered.

Checks should also be made on the condition of valleys, soakers and flashings. Where a roof slope is to be retained, as at the front of the building, it is worth considering re-roofing this at the same time as the conversion. Re-roofing at a later date is likely to lead to damage to internal finishes that are applied as part of the conversion process. Consideration should also be given to the nature of any replacement roofing material, not just on aesthetic grounds but also on loading. Replacing slates with concrete tiles, for example, will increase roof loading and may therefore have structural implications. Where re-roofing takes place, the provision of breather membrane rather than conventional underlay may be considered. Note that, in all cases, roof ventilation should be provided with a 50-mm air gap to the cold side of the insulation, and vents at the eaves and ridge (Fig. 9.20a).

Walls

In a terraced dwelling, the conversion loading is usually transferred via beams to party walls, or to a party wall and a flank wall if at the end of a terrace (see Fig. 6.1 for a typical beam configuration). Note that front and rear walls are often heavily compromised by existing structural openings and are generally not used to support new beams.

Any wall required to support a new load must be thoroughly assessed. Party walls have the disadvantage of being difficult to examine easily, but it is essential nonetheless that the extent of buttressing relative to length, mode of construction and thickness be determined. The same checks must be carried out on flank walls.

In some pre-twentieth century dwellings, a party (compartment) wall separating buildings may not extend into the loft. A check should be made of the adequacy of any party/compartment wall in the roof void. This must include a check on fire-stopping between the wall and roof structure, the presence of perforations or gaps in the masonry between the dwellings, and the thickness.

External load-bearing walls should be visually inspected, noting the method of construction and general condition. Cracked masonry or evidence of bowing or buckling may require further investigation and remedial action. Similarly, mortar condition must be noted, and walls and chimneys repointed where necessary. Note that the strength of brickwork in older buildings where lime mortar has been used (generally pre-1920) may be relatively limited. The strength of both bricks and mortar must be taken into account: an unfactored stress of 0.42 N/mm² is sometimes assumed in these cases.

Gables should be checked for plumb, particularly where it is proposed to extend the gable to form a rear flank gable (Fig. 8.1b). Chimney stacks to be incorporated within a new wall in a hip-to-gable conversion (as illustrated in Fig. 8.10) should be similarly checked. Note that, over time, a combination of sulphate attack and weathering may cause chimney stacks, and sometimes gable walls within which chimney breasts are accommodated, to lean out of the perpendicular. This effect is caused by the expansion of mortar and it is most pronounced on the masonry face that has the greatest exposure to rain (a typical leaning chimney can be seen in Fig. 3.5).

Solid brick walls for dwellings were common until the 1920s and were still being used, sporadically, as late as 1939. An indication may be provided by the bond (generally Flemish or English for solid walls). In cases where solid masonry walls are built with an outer face in stretcher bond, as is sometimes the case, tying between the inner and outer courses may be absent and the stability of the wall must be considered. Solid brick walls are often generically described as 9" (225 mm) walls but note that (a) this is only a nominal designation because brick lengths varied and (b) overall wall thickness was height-dependent (see Table 5.1). 'Brick on edge' and $4^{1}/2$" walls were generally used for internal partitions, but there are examples of party (compartment) walls of $4^{1}/2$" thickness. Nine inch (225 mm) walls were widely used in external and party walls in two-storey dwellings and are commonly encountered. For wall heights of up to 40', combinations of 18", $13^{1}/2$" and 9" were sometimes used. $13^{1}/2$" walls were also used in any dwelling where the height of one storey exceeded 10'.

Cavity walls were adopted after 1920 and were in general use by the outbreak of World War II, although the earliest examples appeared at the beginning of the nineteenth century. External evidence of cavity walls is generally provided by the use of stretcher bond, i.e. a running bond without headers.

Until the early 1970s, cavity walls were constructed with an outer leaf half-a-brick (about 100 mm) in thickness with a 2" (50 mm) cavity width and an inner, generally load-bearing, leaf formed from 4" (100 mm) blockwork. Later variations included the use of lightweight blockwork on the inner leaf, a 75 mm cavity width and the use of cavity insulation, which was routinely provided from the mid-1980s. Inner and outer leaves are tied together with proprietary wall ties.

Table 5.1 Solid brick masonry walls.

Brick fraction	Nominal thickness
Brick on edge	3" (75 mm)
$^{1}/2$ brick	$4^{1}/2$" (112 mm)
1 brick	9" (225 mm)
$1^{1}/2$ brick	$13^{1}/2$" (337 mm)
2 brick	18" (450 mm)

In most cases, the inner leaf of a cavity wall performs the load-bearing function. In earlier masonry-constructed buildings with cavity walls (from the mid-1920s onwards) the inner leaf is generally built from brick or clinker blocks. Lightweight and, latterly, ultra-lightweight blocks have been used in the construction of the inner leaf. It should be noted that ultra-lightweight blocks may have restricted load-bearing potential.

Timber-framed domestic dwellings, with internal softwood wall frames that carry the load of floors and roof, have been constructed in growing numbers since the 1960s. Note that the external brick skin of a timber-framed building is generally not intended to support any of the elements of structure. However, in the case of both timber frame and cavity masonry walls, the outer leaf is designed to resist wind loads.

In all cases (both for timber frame and masonry buildings) it is necessary to check the adequacy of wall plates and, in the case of timber frame buildings, head binders, if it is proposed to use these to support elements of the new structure.

The position of doors, windows and other openings should be recorded as part of the survey. The general condition of wall openings should be examined, and lintels exposed, checked for adequacy and replaced where necessary. This is of importance where walls contain wide openings. Particular attention must be paid to upper-storey lintels where there may only be a few courses of brickwork between a window opening and wall plate. Note that, in many cases, bays have only limited load-bearing potential.

Foundations

Before the widespread availability of mass concrete in the early twentieth century, the footings of most domestic buildings were relatively shallow, with stepped courses of brickwork built directly off the subsoil. A building's foundations and the subsoil supporting them must therefore be considered in relation to the proposed new loading, which is generally increased by about one third when a loft conversion is added to a two-storey dwelling.

A trial pit may be excavated to allow the footings to be checked. The form of the substructure, and particularly the projection (width) of footings, can then be examined and the bearing capacity of the subsoil assessed. Remedial measures, such as underpinning, may be required is some circumstances.

To a certain extent, however, routine conversions in single- and two-storey dwellings represent an exception to this. Below-ground inspections are not always requested by building control bodies: foundations are sometimes assumed to be adequate and this assumption is based to a great extent on the local knowledge of the building control service. In all cases, existing and proposed loads, including the extent of load dispersal at the base of walls, must be considered. In assessing the need for checks on foundations, the following points may be considered:

- Geotechnical factors
- Historical factors
- Additional factors

Geotechnical factors

Knowledge of local subsoil, taking into account bearing capacity, potential for volume change and height of the water table. Foundation movement may be manifested by the presence of movement cracks and movement away from the perpendicular in existing external and internal walls. Conversely, it may be noted that ground conditions beneath foundations in older buildings may improve as a consequence of long-term consolidation.

Historical factors

Including the likely or known configuration of substructure (composition, width and depth of foundations) based on the age and type of dwelling.

Additional factors

History of local subsidence/settlement/heave particularly in clay subsoils; history of remedial works to buildings in the immediate vicinity. History of drainage repairs. Proximity of trees, noting species. Note that the localised removal of trees in clay subsoils may increase the risk of heave.

Foundation movement and damage to dwellings in the UK is most commonly caused by the expansion and contraction of clay subsoils rather than by additional loading above ground. Anecdotal evidence suggests that routine loft conversions in single- and two-storey buildings do not, generally, compromise foundation stability. However, any additional loading has the potential to contribute to foundation movement, particularly if the ground is subject to volume increases or decreases caused by changes in soil moisture levels.

Internal walls and partitions

Internal walls and partitions should be examined. There are two main reasons for doing so. One reason is to ensure (in the case of walls bounding a protected stairway) that the walls are capable of offering an appropriate degree of fire resistance. The other is to identify internal potential load-bearing structure. Before the advent of trussed rafters, a high proportion of traditional 'cut' roofs depended on the provision of an internal spine wall to transfer some roof loads to foundations, as well as to provide support for intermediate floors and, in the roof space, ceiling joists. Where suitable, spine walls are sometimes used to provide support for the floor structure of a conversion.

The only way to assess whether an internal wall is capable of supporting new or different loads is to examine the whole wall from the roof space down to the foundations. The materials used, the method of construction, the vertical continuity between floors (the walls should not be offset between floor levels) and the type of foundations must be noted. It may be necessary to expose the footings and fabric of such a wall to confirm its load-bearing potential.

In buildings constructed after 1920, internal load-bearing walls may be formed from brick, block or timber studwork. Earlier buildings may use brick or studwork with a lath and plaster finish, and sometimes studwork with brick infill. Original drawings of the building should be checked if these are available. In many cases it will be necessary to determine the load-bearing suitability of internal walls by calculation. This will certainly be the case if a spine wall has been altered. This would apply, for example, where two rooms have been knocked together at ground floor level and, in this case, it would also be necessary to check the adequacy of the beam provided.

A traditionally-constructed internal stud wall built from 100 × 50 mm studs at 400 mm centres with 16 mm lath and plaster to both sides might provide 30-minute fire resistance without modification if the lath and plaster is in good condition.

Doors between habitable rooms and the proposed protected stairway should also be examined. Conventionally glazed doors and fanlights to the enclosure are generally not acceptable because they do not offer adequate fire resistance. Conventional glazing is usually acceptable in final exit doors, however (but see also Chapter 4).

Floor and ceiling structure

All existing floors and ceilings must offer an appropriate level of fire resistance (see Chapter 4). In addition, the floor structure of the existing upper floor of the building must be capable, in most cases, of providing support for a new staircase to the conversion.

The fire resistance of a floor is dependent on a number of factors. These include the composition and condition of the ceiling, the size and spacing of joists and the type and condition of floorboards. In addition, the presence of gaps between the floorboards of intermediate floors should be noted because these compromise fire resistance. This applies particularly with plain edge boards, but it is also relevant with tongued and grooved boards that have shrunk (see also Figs 7.17 and 7.18).

Strength of existing timber elements

The introduction of strength grades for timber is a relatively recent development. The first grading systems emerged in the late nineteenth and early twentieth centuries but these were somewhat haphazard and it was not until the immediate post-war years that formalised approaches to testing and grade marking were developed. Without conducting tests, it is not possible to know the strength of existing timber elements in older buildings. Assumptions are sometimes made on the basis of the species of timber used and this may be ascertained by an experienced observer.

Water tanks

The position, dimensions, height above floor level and functions of water tanks must be noted. In most cases, it will be necessary to move the tanks to make the best use of the available floor space. The position of pipes must also be noted as part of the survey.

Cold water storage tank

This generally serves a dual function. In a conventionally-plumbed dwelling, it feeds water to lavatory cisterns and cold tap outlets (but generally not the cold tap in the kitchen). Where vented hot water cylinders are used, the cold water tank also provides a cold feed to the hot water cylinder and a termination for the hot water cylinder vent pipe.

The existing tank can be moved, if there is space to accommodate it. Common new positions include the eaves and, where headroom permits, the head of the new access stair or the apex of the roof. In cases where the existing tank cannot be accommodated in this way, installation of a space-saving 'coffin' tank should be considered (see Glossary). Because the head of water (pressure) at tap outlets is a function of relative tank height in conventional systems, any new tank position that is lower than the original must be carefully considered. An alternative, which is sometimes appropriate when the conversion is being carried out as part of a wider renovation, is to eliminate the need for a cold-water tank by changing to an unvented hot water system (also known as a Megaflo system). Under this arrangement, the hot water system operates at mains pressure. In addition, cold taps and cisterns previously fed by gravity are switched to the mains feed.

Where an unvented hot water system is proposed, it is essential that a flow and pressure test be conducted as part of the survey to establish that the mains supply meets the unvented system's flow and pressure criteria. Most water companies aim to supply water at 1 bar, which is the minimum pressure required for most unvented systems. However, the minimum pressure water companies are obliged to maintain is 0.7 bar. Note that building control must be notified where unvented systems are proposed and that such systems must be commissioned and certified by the installer.

Central heating feed and expansion (F&E) cistern

This is an integral part of open-vented central heating systems. It keeps the system topped up and accommodates excess water created by system overheating. In order to eliminate the need for such a tank, a closed or sealed central heating system would need to be installed.

Drainage and services

The survey should include an accurate record of the existing arrangement of drainage and other services. The position of the soil and vent pipe is often of critical importance where the conversion is to be provided with sanitary facilities. Achieving seamless integration between old and new systems is frequently troublesome in loft conversions. The position of passive vent ducts must also be recorded. Again, in terms of integrating old and new, attention should also be given to the position and adequacy of the existing rainwater drainage system and guttering. Similarly, the position and orientation of ancillary elements such as aerials should be considered if there is likely to be a need to relocate them.

Chimneys

Externally, the position of flues and chimneys relative to the proposed conversion and its windows should be noted and measured. The presence or otherwise of chimney cowls should also be recorded.

The position of chimney breasts within the dwelling must also be recorded in the survey. This is often a critical factor in determining the arrangement of the conversion, particularly where a steel roof beam is required to span the width of the building. It is generally not permissible to fix structural members, such as beams and joists, to a chimney breast.

The path of the chimney breast should be tracked through the building and, where chimney breasts have been removed, it is prudent to check that the remaining masonry is adequately supported. In some cases, it will be necessary to retrospectively install a means of support. Corbelled brickwork or gallows brackets (see Glossary) are sometimes acceptable for this purpose although it may be necessary to install a supporting beam spanning between walls.

6 Beams and primary structure

When a loft is converted, new loads must be transferred to the foundations of the existing building. These additional loads may be channelled via beams to external walls and sometimes to internal load-bearing walls. The purpose of this chapter is to consider the characteristics, use and methods of fixing for both steel and timber beams.

Beams and other critical structural configurations should be designed by a structural engineer or other competent person.

STEEL AND TIMBER STRUCTURAL DESIGN CODES

Table 6.1 is intended to clarify the status of design codes and standards as they stand at the time of writing. Note that the basis of structural design is in a state of transition between traditional permissible stress and newer limit state principles. Although limit state principles are now in general use for structural steel, they have yet to be widely adopted for structural timber design.

Note that BS 449-2: 1969 *Specification for the use of structural steel in building* is no longer referenced in Approved Document A *Structure* (2004) but it continues for the time being as a British Standard. The Approved Document now refers to BS 5950 (*Structural use of steelwork in building*) which is based on limit state principles. However, the Approved Document references the BS 5268 series (*Structural use of timber*) which is based on permissible stress.

One of the arguments advanced in support of continuing to use BS 449, for relatively modest domestic construction at least, concerns the intermixing of timber and steel structural elements. A common example of this in loft conversions is when a timber post is used to support a steel ridge beam (Fig. 6.1). In the case of the timber post and steel ridge beam, it would be necessary to combine a limit

Table 6.1 Structural design of timber and steel.

Standard series or code	Design application	Basis	Status in AD A *Structure* (2004)	General status (August 2005)
BS 449	Steel	Permissible stress	No longer included	Current
BS 5950	Steel	Limit state	Included	Current
Eurocode 3	Steel	Limit state	Not yet included	Current/national annexes pending
BS 5268	Timber	Permissible stress	Included	Current
Eurocode 5	Timber	Limit state	Not yet included	Current/national annexes pending

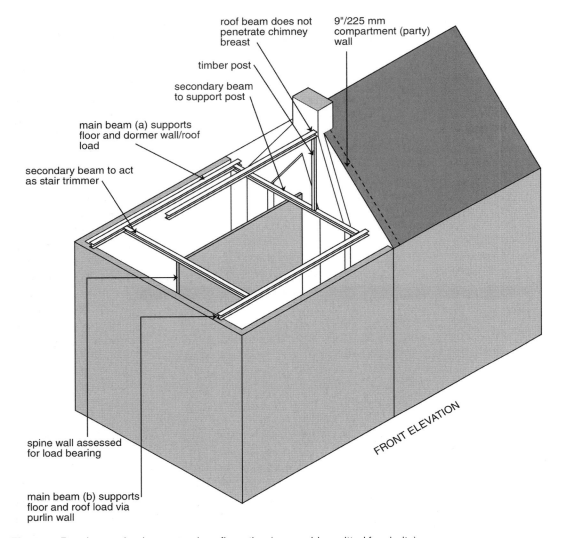

Fig. 6.1 Box dormer: basic structural configuration (near gable omitted for clarity).

state calculation (for the steel beam) with a permissible stress calculation (for the timber post).

BEAM POSITION RELATIVE TO EXISTING STRUCTURE

Headroom is a valuable commodity in most loft conversions. One of the most effective methods for preserving headroom is to limit the depth of the floor structure. In order to do this, it is generally necessary to limit the clear span of joists to about 4 m between supports. For a number of reasons, including the joist depth required, it is generally not feasible to exceed this by much. As most buildings are considerably deeper than 4 m, support for joists, either at one or both ends, must be provided by beams. Note that the maximum span of any floor supported by a wall is 6 m (Approved Document A *Structure*, 2C23).

In loft conversions, it is practice to support the main beams in flank or compartment walls (i.e. generally the side walls of the building). Flank walls generally contain few openings likely to prejudice the stability of the wall (none in the case of a mid-terrace house). This should be contrasted with the front and rear walls in most dwellings which characteristically feature numerous openings, often with lintels of unknown strength. In addition, ceiling joists often run parallel to flank walls, and fixing beams at right angles to these allows new floor joists to be accommodated between them.

In most cases, universal columns or universal beams are used in loft conversions. Steel beams are widely available and, because they have a small sectional size relative to their strength, they can be positioned more easily relative to existing elements of structure. Timber composite and laminated beams may also be used, although their relative size means they are somewhat impractical for use in conversions. Flitch beams, however, are used relatively widely, often as floor trimmers and in certain roof beam applications. Characteristics of both steel and timber composite beams are outlined below. In all cases, a clearance of 25 mm should be provided between floor beams and existing elements (such as lower-floor ceiling joists) in order to minimise the risk of damage to retained structure caused by deflection.

BEAM CHARACTERISTICS

Beams in loft conversions often serve more than one purpose (Figs. 6.1 and 6.2). Where it is positioned towards the front of the dwelling a floor beam (sometimes also called a foot beam) may also provide support for the front roof slope via a purlin wall). Similarly, as well as providing support for the flat roof of a new dormer, a roof or ridge beam may also provide support for an existing front roof slope. It is important to note that a roof beam supported at one end by a timber post will itself require lateral restraint.

Subsidiary beams supported by a pair of main beams (for example, the stair trimmer and timber post support in Fig. 6.1) are generally described as secondary beams.

Common structural steel sections

Hot-rolled steel sections are widely used as beams in the process of loft conversion. Steel has the advantage of being relatively inexpensive. It is also widely used and therefore readily accepted by local authority building control when appropriate calculations are provided. The following sections are commonly used to provide support for new floor, wall and roof structures in conversions:

- Universal beams
- Universal columns

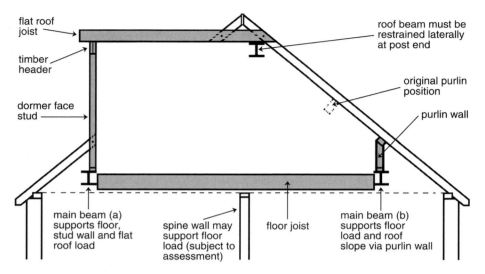

Fig. 6.2   Box dormer: common configuration of elements.

- Parallel flange channels
- Rolled steel joists

Universal beam (UB)

Universal beams are produced in a range of standard sections designated by depth, width and mass. Thus a universal beam designated $203 \times 133 \times 25$ has a depth of 203 mm, a width of 133 mm and a linear mass of 25 kg/m (nominal). In all cases, the depth of a universal beam is greater than its width. The flanges of a universal beam, both internal and external, are parallel (Fig. 6.3a).

Universal column (UC)

Used both in vertical and horizontal configurations. Designated in the same way as universal beams by nominal depth, width and linear mass. Flanges are parallel inside and out. Widely used as floor beams where it is necessary to accommodate the beam within the depth of the floor structure (Fig. 6.3b).

Parallel flange channel (PFC)

Designated by depth, width and linear mass. Typical applications include the replacement of timber purlins (Fig. 6.3c).

Rolled steel joist (RSJ)

The term RSJ is often applied generically to steel beams. In strict terms, this is incorrect because RSJ is a designation in its own right. RSJs were largely super-seded in the construction industry by the introduction of universal beams and

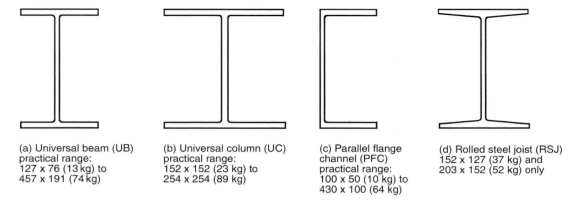

(a) Universal beam (UB)
practical range:
127 x 76 (13 kg) to
457 x 191 (74 kg)

(b) Universal column (UC)
practical range:
152 x 152 (23 kg) to
254 x 254 (89 kg)

(c) Parallel flange
channel (PFC)
practical range:
100 x 50 (10 kg) to
430 x 100 (64 kg)

(d) Rolled steel joist (RSJ)
152 x 127 (37 kg) and
203 x 152 (52 kg) only

Fig. 6.3 Structural steel sections.

columns in the late 1950s. RSJs have tapered internal flanges and are therefore less convenient when inserting bearers or the ends of joists into the webbing. RSJs are still produced in very small numbers (predominantly for the mining industry) and find occasional use (Fig. 6.3d).

Engineered timber beams

The depth of section required means that solid timber members are generally not practical in longer beam applications. Even in the context of joists, solid timber is generally limited to spans of about 4 m. Engineered timber beams, which may take the form of laminated solid sections or fabricated composite beams, provide considerably more scope, and the characteristics of some common forms are outlined below:

- Flitch beams
- Laminated timber beams
- Timber composite beams

Flitch beams

Flitch beams (sometimes called sandwich beams) are widely used in loft conversions (Fig. 6.4). The beam comprises a central steel plate generally between 8 and 20 mm thick sandwiched between strength-graded timber sections. The flitch is the only composite beam that can be site-fabricated. As with other beams, the use of flitch beams is subject to the provision of structural engineering calculations. However, as a rule of thumb, a flitch beam is equivalent to a solid timber section approximately 30 times the thickness of its steel plate. Bearing plates or padstones are generally required at supported ends. Flitch beams offer relatively little resistance to lateral forces and so restraint is generally provided by joists positively fixed to the side of the beam. Beam components are generally fastened to each other by bolting through at intervals of not more than 500 mm. Note that work

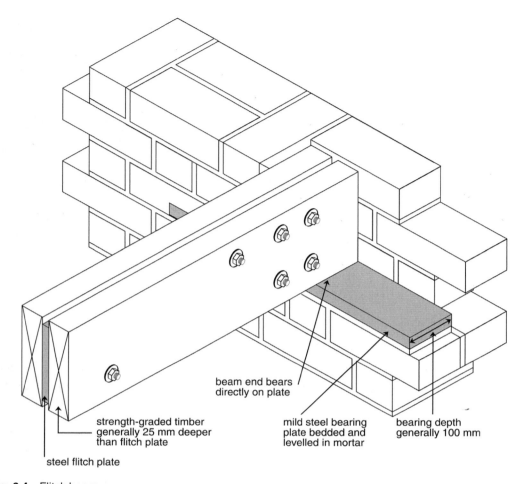

strength-graded timber
generally 25 mm deeper
than flitch plate

steel flitch plate

beam end bears
directly on plate

mild steel bearing
plate bedded and
levelled in mortar

bearing depth
generally 100 mm

Fig. 6.4 Flitch beam.

on developing a 'new age' flitch, which does not depend on bolt fastenings, is currently being undertaken.

Laminated timber beams

A number of proprietary versions are available. These include: LVL (laminated veneer lumber, Fig. 6.5a) which has relatively slender vertical laminations; and glued laminated timber (glulam, Fig. 6.5b) which is characterised by horizontal solid timber laminations with a depth of 45 mm. Note that glulam has certain aesthetic benefits. Planed and treated, it may be left exposed.

Because they are of constant section, laminated timber beams may be cut to length without fear of compromising their structural integrity. The disadvantage is that relatively deep sections are required. Fixed at low levels, laminated beams would block access to an eaves space; at higher levels, beams of this sort would clearly have an impact on available headroom.

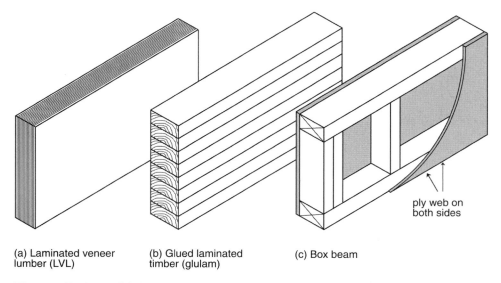

(a) Laminated veneer lumber (LVL) (b) Glued laminated timber (glulam) (c) Box beam

Fig. 6.5 Engineered timber beams.

Timber composite beams

Timber composite beams may be used as purlins or sometimes to provide floor support (Fig. 6.5c). Composite beams generally incorporate softwood flanges top and bottom with deep plywood webs. Beams of this sort have the advantage of offering a high strength-to-weight ratio.

Given their considerable depth and the associated handling problems, composite beams are perhaps best suited to situations where a large roof is being replaced or in new-build applications. The integrity of such beams is usually dependent on gluing, and beams are generally manufactured by specialists under controlled conditions – site fabrication is generally not possible. Equally, because they often depend on correctly spaced and prefabricated internal stiffeners, composite beams cannot generally be cut to length without appropriate modification.

FIRE RESISTANCE OF BEAMS

Beams supporting a new floor require a minimum 30-minute fire resistance. Note that a roof beam would generally not require 30-minute fire resistance. Approved Document B *Fire* (section 8.4) excludes a structure that only supports a roof unless, for example, the roof formed part of an escape route or was essential for the stability an external wall that needs to have fire resistance.

Protection of structural elements such as floor beams may be provided in a number of ways:

■ *30-minute fire-resisting ceiling below beam*: an imperforate ceiling of appropriate construction may be provided beneath the beam (see Fig. 7.3).

■ *Boxing-in*: in cases where a steel beam supports a floor from beneath, it is possible to provide protection by cladding all exposed sides of the beam with an appropriate fire-resisting board (see Fig. 7.3).

■ *Intumescent paint*: this is a practical consideration if a steel beam is providing support from below. It is not as effective a means of protection if joists or other elements are fixed into the webbing because the paint requires a space into which to char in order to provide protection.

Consideration should also be given to the fire resistance of the supported ends of beams. This would apply, for example, where a beam is supported by a compartment wall separating buildings (a party wall). In general, a bearing of 100 mm is provided, and in a 9" (225 mm) wall this would generally mean that the beam end was protected by masonry at least 100 mm thick (i.e. half a brick) on the adjoining owner's side. This is usually sufficient, provided any gaps are sealed. Additional protection measures may be necessary where end cover is less than 100 mm, or where the depth of cover is not known.

BEAM BEARINGS

Beams are supported at right angles to their span. However, supporting a structural steel beam or flitch beam by allowing it to bear directly onto bricks or blocks is not usually considered to be satisfactory. Concentrated loads at supported ends will create bearing stresses that may exceed permissible values and this can lead to localised failure in the masonry which may take the form of spalling or cracking. In many cases, the bearing strength of existing masonry is not known and therefore a conservative value must be adopted. As noted in Chapter 5, a value of $0.42 \, \text{N/mm}^2$ is sometimes assumed. However, the overall compressive strength of masonry walls constructed from soft stock bricks bedded in lime mortar may be as low as $0.21 \, \text{N/mm}^2$ (BS CP 111 part 2: 1970).

To allow a broader distribution of the load on the wall, it is normal practice for the structural engineer to specify the use of an intermediate element of known strength. This generally takes the form of a mild-steel bearing plate or a padstone.

Beams should bear at full width on any padstone or bearing plate. Note that any proposal to notch or chamfer a beam end (for example, where it must be fitted beneath a roof slope) should be supported by a structural engineer's calculations. It is not always possible to cut a beam in this way. In these cases, it is necessary to allow the beam to project beyond the roof slope, where it may be protected by lead cladding.

Mild steel bearing plates

Where steel beams, especially floor beams, are to be installed as part of a loft conversion, mild steel bearing plates are often used in preference to padstones because there is less need to remove masonry (Fig. 6.6).

Mild steel bearing plates used in loft conversions are available in a range of standard thicknesses and are cut to length by suppliers. In most cases, calculations

Table 6.2 Common sizes
of concrete padstones (mm).

215 × 140 × 102
300 × 140 × 102
440 × 140 × 102
215 × 140 × 215
440 × 140 × 215
215 × 215 × 102
440 × 215 × 102

are based on a bearing depth of 100 mm, which co-ordinates with the thickness of a single brick or blockwork leaf. Bearing plates are bedded and levelled in mortar. Note that the beam end should bear directly on the plate with no intermediate use of mortar. Short beam sections (sometimes called cuttings) may be used in this application, subject to calculation.

Padstones

Padstones can be cast in situ, cast in moulds on site, or purchased from suppliers in various sizes. Table 6.2 indicates the range of padstones produced by a typical manufacturer. However, off-the-shelf padstones are sometimes difficult to source to a specific size.

Pre-cast padstones are produced under controlled conditions using concrete that is compacted by vibration to increase density before being allowed to cure. This has the advantage of producing a component of known compressive strength – typically in excess of $50 \, N/mm^2$ 28 days after manufacture. As Table 6.2 indicates, factory-produced padstones are generally produced in sizes designed to co-ordinate with standard bricks and blocks. Factory-produced padstones generally do not contain reinforcing bars.

Padstones are fixed with mortar in recesses created in the masonry (Fig. 6.7). It should be stressed that steel beams must bear directly on the padstone with no intermediate use of mortar.

Casting padstones on site, whether in situ or otherwise, can be a less costly alternative to purchasing them. However, unless the mix is tested, the padstones produced will not be of known compressive strength. Attention should also be given to the amount of time needed for the concrete to cure before any beam is fixed and loaded. To limit the risk of shear failure within the concrete when padstones are cast on site without reinforcing bars, the depth of the padstone should exceed the offset (Fig. 6.8).

Pre-cast concrete lintels of appropriate dimensions and compressive strength are sometimes used as an alternative to padstones. However, the presence of steel reinforcing bars in concrete lintels should be considered in relation to any subsequent requirement to drill and fix to them.

In order to accommodate a padstone, a considerable depth of brickwork must be removed. Where floor beams are to be installed close to existing ceiling joists, it

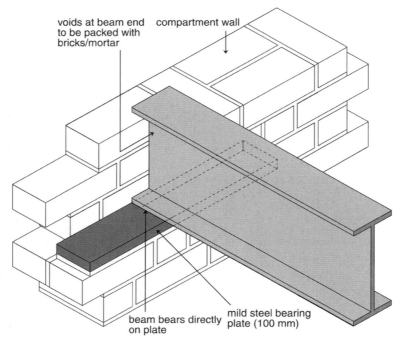

voids at beam end
to be packed with
bricks/mortar

compartment wall

beam bears directly
on plate

mild steel bearing
plate (100 mm)

Fig. 6.6   Beam support: mild steel bearing plate.

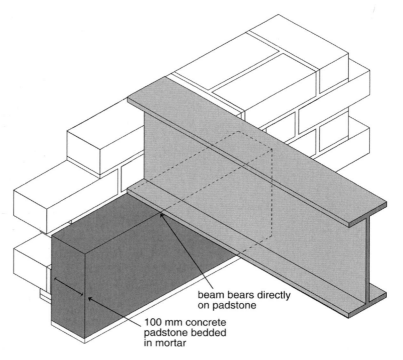

beam bears directly
on padstone

100 mm concrete
padstone bedded
in mortar

Fig. 6.7   Beam support: concrete padstone.

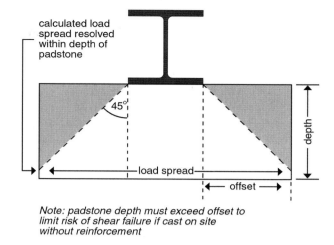

Note: padstone depth must exceed offset to
limit risk of shear failure if cast on site
without reinforcement

Fig. 6.8 Site-cast concrete padstone.

will be difficult to get access to the masonry to achieve this, and there is a risk of damaging walls and finishes in the rooms below.

BEAM PENETRATION

As noted above, beams must be adequately supported by padstones or bearing plates. However, consideration must be given to the extent of beam penetration into walls. Typically, beams for loft conversions are designed with a notional end bearing of 100 mm. This reflects the fact that the inner (load-bearing) leaf of most cavity walls is 100 mm. Equally, in a 9" (225 mm) wall, removing brickwork to a depth of half a brick would yield a similar depth of bearing. In most cases, the specification of a 100-mm depth bearing means that the supported end of the steel does not extend beyond the notional centreline of a 9" party wall, and this is clearly of benefit to an adjoining owner who may wish to conduct a similar conversion (see also Chapter 1, Party Wall Act and Chapter 5, External relationships).

It is generally not possible to use a chimney breast to support a beam (guidance on fixing near flues is provided in Approved Document J *Combustion appliances and fuel storage systems*). This often leads to difficulties where it is necessary to install a beam at ridge height. A commonly adopted approach in these situations is to support the chimney end of the ridge beam on a timber post, generally 100 mm square in section and fixed top and bottom in fabricated mild steel shoes. The upper shoe is generally bolted to the flange of the ridge beam while the lower shoe is supported either by a suitable internal load-bearing wall, a spine wall for example, or by a secondary steel beam running parallel to the chimney breast (Fig. 6.1). Note that the post-supported end of the beam must be restrained to prevent lateral movement.

BEAM SPLICES

Manoeuvring a steel beam that is usually at least 200 mm longer than the gap it is intended to span, and in the confines of a roof space, is a troublesome matter, particularly where there are obstacles such as existing elements of roof structure, chimney breasts and water tanks. In semi-detached and end-terrace houses, it may be possible to create an opening though a gable wall or roof hip to facilitate final positioning. In terraced houses, however, it is generally considered prudent to avoid fully penetrating the compartment wall. Sometimes an arc is chased into the brickwork to allow the beam to be moved into its final position.

An alternative approach, and one that makes handling the steel in situ considerably easier, is for the beam to be cut into two or more sections and for spliced connections to be provided. There are three principal methods of splicing beams in these situations. In all cases, the design of the beam and its connections, including the specification and number of bolts, is determined by the structural engineer.

Flange and web plate splice

This is the most commonly used engineer-specified splicing system (Fig. 6.9). High-tensile bolt assemblies are used with flange and web plates to form connections between the beam sections.

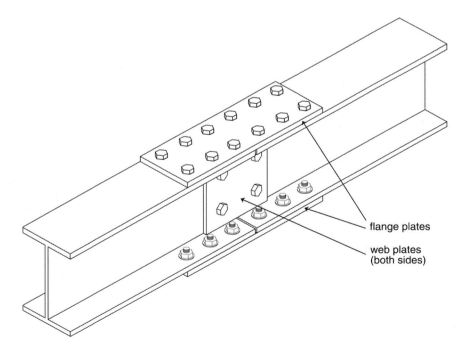

flange plates

web plates
(both sides)

Fig. 6.9   Flange and web plate splice.

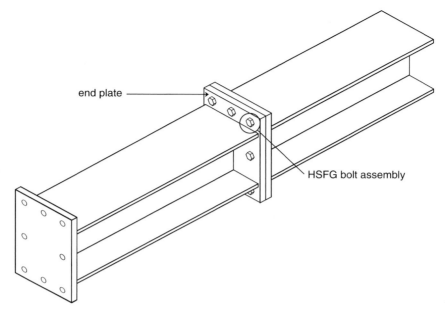

Fig. 6.10a End plate beam splice.

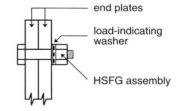

Fig. 6.10b HSFG assembly.

End plate beam splice

Beams are provided with end plates that are machined and shop-welded to the beam ends (Fig. 6.10a). A site-bolted connection is made between the adjacent end plates. This approach is more expensive than conventional flange and web splicing, and installation must be carried out with great care: the integrity of the splice is dependent on the use of high-strength friction grip (HSFG) bolts (see *Bolted connections* below) and these must be correctly tensioned (Fig. 6.10b).

The disadvantage with splicing beams using either system is that there are inevitably bolt and plate projections above and below the beam flanges. Any such projections should be considered relative to ceiling heights and floor clearances.

PFC bearing

With this arrangement, the main beam itself does not penetrate the supporting walls: final support is provided by a pair of parallel flange channels at each end

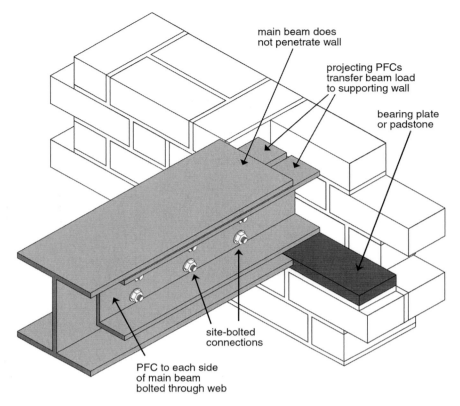

main beam does
not penetrate wall

projecting PFCs
transfer beam load
to supporting wall

bearing plate
or padstone

site-bolted
connections

PFC to each side
of main beam
bolted through web

Fig. 6.11a   PFC bearing.

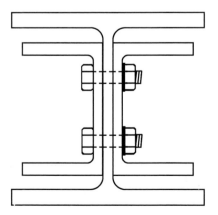

Fig. 6.11b   PFC bearing – section.

(Fig. 6.11). These are accommodated within the web of the main beam. This approach eliminates some of the problems traditionally associated with positioning beams between existing masonry walls.

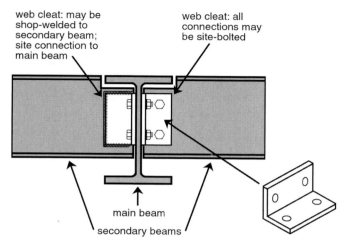

web cleat: may be
shop-welded to
secondary beam;
site connection to
main beam

web cleat: all
connections may
be site-bolted

main beam

secondary beams

Fig. 6.12 Cleat connections.

Beam-to-beam connections

Steel beams may be joined at right angles by means of cleat connections. The design and specification of such connections is, of course, the role of the structural engineer. Web cleats may be shop-welded to both sides of the web of the secondary beam. In addition, it is usually necessary for one of the flanges of the secondary beam (generally the upper flange) to be notched at the joint in order to achieve a level surface between it and the main beam. Site connection is achieved by bolting (Fig. 6.12).

BOLTED CONNECTIONS

Bolted connections are used in a range of structural applications. At one end of the scale, mild steel bolts may be used in the fabrication of simple timber-to-timber components such as trimmers and header beams. In situations such as these, the strength of bolts is unlikely to be a critical factor provided they are of adequate dimensions and appropriately spaced.

The opposite end of the spectrum is represented by bolted connections in beam splices. For example, the design of bolted connections in an end plate beam splice (described above) would include specification of high strength friction grip (HSFG) bolt assemblies. It should be noted that proof loads for HSFG assemblies are between two and three times greater than those for black (mild steel) bolts. However, HSFG and black bolt assemblies are similar in appearance, and the ability to correctly identify them is therefore of considerable importance. The following notes are provided to clarify the use of various common bolt configurations.

It is important to note that the finish or colour of a bolt does not provide a reliable indication of its strength or function. 'Black' bolts, generally the weakest used in structural connections, are not always black. Conversely, high-strength friction grip bolts (the strongest) may have a black phosphate finish.

Grade 4.6 bolts

Grade 4.6 bolts and nuts manufactured from mild steel are often referred to as 'normal strength' or 'black' bolts. Grade 4.6 bolt assemblies may be used in timber-to-timber and timber-to-steel connections (Table 6.3).

It is necessary to use washers both behind the bolt head and behind the nut when making timber connections. In addition, it is often necessary to use toothed plate connectors between timber elements (see notes below). The diameter of washers in timber connections should be three times bolt diameter and the thickness of washers 0.25 of bolt diameter. Clearance holes are generally diameter + 2 mm. Thus an M10 bolt would require a 12 mm clearance hole and 30 mm washers. Note that, when tightened, at least one complete thread should protrude beyond the nut.

Table 6.3   Grade 4.6 bolt assemblies.

Identification	Bolt head stamped 4.6
Material	Mild steel
Applications	Timber-to-timber and timber-to-steel
Clearance hole	Generally bolt diameter + 2 mm
Proof load (M16)	34.8 kN

Grade 8.8 'high strength' bolts

These are sometimes described as high strength or hexagon structural bolts (Fig. 6.13a). They are manufactured from high-tensile steel and are used in a wide range of structural applications including steel-to-steel connections (Table 6.4). Note that high strength bolts are not the same as high strength friction grip (HSFG) bolts.

Table 6.4   Grade 8.8 assemblies.

Identification	Bolt head stamped 8.8
Material	High tensile steel
Applications	Timber-to-timber, timber-to-steel, steel-to-steel
Clearance hole	Design dependent
Proof load (M16)	89.6 kN

High strength friction grip (HSFG) assemblies

These bolt assemblies are used in structural steel joints where high clamp forces are required to generate friction between mating surfaces (Fig. 6.13b). A typical application would be in the formation of bolted joints in beam splices where beam sections are connected via end plates. Like other high-strength bolts, HSFG bolt

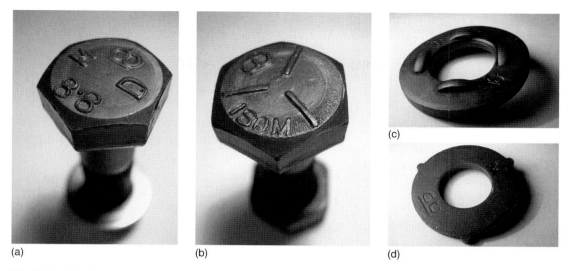

Fig. 6.13 (a) Grade 8.8 bolt. (b) HSFG bolt. (c) HSFG load-indicating washer. (d) HSFG flat washer.

Table 6.5 HSFG assemblies.

Identification	*Bolt head*: three radial lines at 120 degrees
	Nut: three circumferential arcs
Material	High tensile steel
Applications	Steel-to-steel connections
Clearance	Design dependent
Proof load (M16)	92.1 kN

assemblies are produced from high tensile steel (Table 6.5). However, HSFG and conventional 'high strength' bolts should not be confused: HSFG bolt heads and nuts are wider across flats than grade 8.8 bolts of the same diameter.

Shank tension is critical in applications where HSFG assemblies are used, and load indicating washers (direct tension indicators) are commonly used to ensure that minimum specified shank tension is achieved in the process of assembly (Fig. 6.13c). These washers have protrusions that are flattened progressively as shank tension increases to pre-determined levels. Non-indicating flat washers for use in HSFG assemblies are identified by nibs (Fig. 6.13d).

Toothed plate connectors

Toothed plate connectors, sometimes called timber connector plates, are used in structural timber-to-timber connections where two parallel lengths of timber are required to act in unison. Toothed plate connectors increase the effectiveness of timber joints by improving stress distribution within the joint and by reducing the tendency of timber to shear parallel to the grain when loaded. Common applications include the formation of trimmers, trimming joists and double joists for partition or bath support (see Fig. 7.8).

TIMBER TO MASONRY CONNECTIONS

It is frequently necessary to fix new timber elements such as sole plates or wall plates to existing masonry walls as part of a loft conversion. As with any structural connection, the choice of fixing system is influenced by the loads to which the new elements will be subjected.

The condition of the masonry substrate must also be taken into consideration when selecting a fixing system. The poor mechanical properties of some traditional stock bricks and the relative weakness of lime-based mortars will influence the choice of anchoring system. The effectiveness of the chosen anchoring system is also influenced by factors including embedment depth, edge distance, the nature of the bond between masonry and fixing, and to a lesser extent the shear/tension characteristics of the fixing itself. Timber to masonry anchoring methods include:

- Tension straps
- Expansion bolts
- Chemical anchoring

Tension straps

Generally in 5 mm (heavy duty) or 2.5 mm (light duty) galvanised steel (Fig. 6.14a). Standard width 30 mm, lengths usually from 100 mm to 3600 mm in 100 mm increments. Tension or restraint straps are used for fixing wall plates and may also be used for fixing sole plates to wall heads. The disadvantage with using straps at wall plate level in loft conversions is that the leg of the strap (unless of minimal length) would generally extend through the existing ceiling into the room below and would consequently disturb wall finishes.

Expansion bolts

Expansion bolts provide an effective method of fixing timber to masonry, but only if the masonry is in good condition (Fig. 6.14b). Friable stock bricks with weak traditional mortar are not a satisfactory substrate. Equally, expansion bolts are best used away from the edges of masonry where expansion stress can cause brickwork to fail locally.

Chemical anchoring

Chemical anchoring systems provide, in effect, a glued connection, and this approach overcomes some of the problems associated with mechanical expansion fixings (Fig. 6.14c). A clearance hole is drilled in the masonry, dust removed from the hole, and the reagents are introduced. Proprietary anchors are available, but threaded bar (properly degreased) may also be used to create a stud. This system is well suited to applications where fixings are to be provided close to the edge of masonry walls.

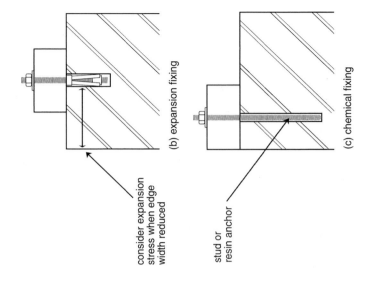

consider expansion
stress when edge
width reduced

(b) expansion fixing

stud or
resin anchor

(c) chemical fixing

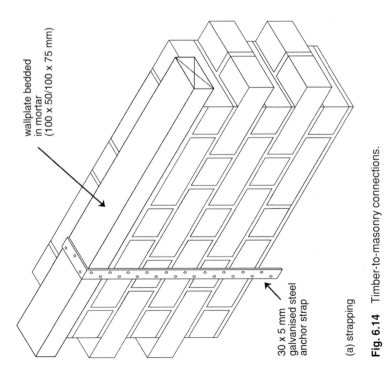

wallplate bedded
in mortar
(100 x 50/100 x 75 mm)

30 x 5 mm
galvanised steel
anchor strap

(a) strapping

Fig. 6.14 Timber-to-masonry connections.

Two types of chemical anchoring are used:

- resin anchors (generally based on polyester resin)
- epoxy resin anchoring systems (which do not expand and may be used in conjunction with stainless steel studs)

DISPROPORTIONATE COLLAPSE

Requirement A3 of the Building Regulations is that buildings '. . . shall be constructed so that in the event of an accident the building will not suffer collapse to an extent disproportionate to the cause.'

Guidance concerning disproportionate collapse is set out in Approved Document A *Structure* (2004) and compliance is primarily achieved by providing effective connections between walls and floors. Specific provisions are dependent on Building Class and this, as far as dwellings are concerned, is defined in relation to the number of storeys in the building and the type of occupancy. Houses not exceeding four storeys are included in Class 1; single occupancy houses of five storeys are included in Class 2A.

The most common sort of loft conversion (one that creates a new storey in an existing two-storey house) would generally not trigger the need for any remedial action elsewhere in the building under the guidance on disproportionate collapse because the building would remain in the Class 1 designation. The same would also apply, usually, to an existing three-storey single occupancy dwelling that becomes a four-storey dwelling as a consequence of a loft conversion. The conversion, of course, would need to conform to Class 1 robustness.

However, when building work creates a five-storey house (i.e. the conversion of the roof space in an existing four-storey house), the Building Class of the entire structure changes from 1 to 2A. In some cases, this might lead to the building being less satisfactory in relation to Requirement A3 than it was before alteration, and the need for remedial action may be triggered. Attention is drawn to the following points.

Basement storeys

Unequivocal guidance on whether basement storeys should be included in the storey total is not provided in Approved Document A. Fig. 6.15 indicates a configuration in which a basement would generally be considered to constitute a storey for the purposes of the guidance. Note that while a basement storey *might* be excluded from consideration if it conforms to certain robustness criteria, this is unlikely to be the case in the mode of construction indicated here.

Status of loft conversions

Requirement A3 (2004) applies to all buildings without limitation and a loft conversion is now considered to constitute a storey in its own right.

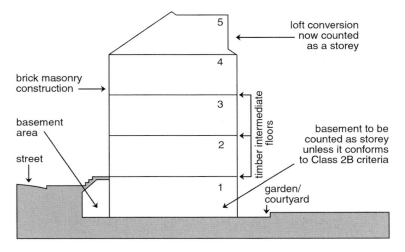

Fig. 6.15   Five-storey dwelling: common configuration.

Change of Building Class – remedial measures for 2A dwellings

Conversions that lead to the creation of a five-storey building are most likely in urban terraced dwellings constructed in the eighteenth and nineteenth centuries. Houses of this sort, constructed from either brick or stone masonry, are generally found only in the larger British towns and cities. As a rule, systematic anchorage between walls and floors was not provided in buildings of the period. Ties between walls and floor joists running parallel to them are likely to be absent; the effectiveness of existing 'natural' ties formed by wall-supported timber joists will vary depending on the nature of the junction and the condition of both masonry and timber.

The implication of the guidance is that Class 2A standards should be applied to the *entire* building when a loft conversion leads to the creation of a five-storey dwelling. However, the Approved Document provides little practical detail and refers readers to additional guidance in BS5628: Part 1 – *Structural use of unreinforced masonry* and BS5950: Part 1 – *Structural use of steelwork in building*.

This would require the retroactive installation of an anchoring system capable of conforming to new-build standards, with straps at 1250 mm spacing between suspended floors and walls (including internal walls) throughout the building. Note that an alternative approach, based on a collapse-resistant support system for the conversion only, is sometimes acceptable.

7 Floor structure

Ceiling joists are seldom capable of supporting occupational floor loads. When a loft is converted, therefore, it is usually necessary to construct an entirely new floor with joists supported by beams or by existing load-bearing walls (see also Chapter 6).

The floor has two primary structural functions. It must first be able to support its own dead load and other designed loads. In addition, it must also be able to support the loads imposed on it by users of the building and their possessions without undue deflection.

Ordinarily, determining loads for timber intermediate floors in dwellings, other than loft floors, is generally straightforward. Unless concentrated loads (caused by trimming members, for example) or line loads (caused by partitions of more than 0.8 kN per linear metre) are factors, there is generally no requirement to provide structural engineering calculations for the joists themselves. Selection of joists appropriate to the loading conditions is made by reference to *Span Tables for Solid Timber Members in Floors, Ceilings and Roofs for Dwellings* (published by TRADA Technology Ltd). These tables, which are referenced in Approved Document A *Structure* (2004) may be used for single occupancy dwellings of up to three storeys.

ROLE OF THE CONVERSION FLOOR

Loft conversion floor designs are generally more complex than those of conventional intermediate floors. There are two reasons for this. One is the need to provide support for the new floor structure, and this often requires the use of structural elements such as beams, the specifications of which must always be justified by calculation.

The other element of complexity concerns the loading to which the new floor may be subjected. Depending on the arrangement of the conversion, the design of the new floor may need to take into account a number of structural loads additional to normal dead and imposed loads. These might include:

- *Roof loads*: for example, where existing purlins are removed and a new purlin wall transmits a proportion of the roof load to the new floor structure.
- *External wall loads*: for example, the loads from new external stud walls including any portion of the roof supported by them.
- *Lower floor ceiling loads*: where ceiling binders are removed, the new floor must provide intermediate support for the ceiling structure in the rooms below.

In cases where roof and wall loads will be supported by the floor joists, rather than directly by elements of structure such as beams or existing masonry walls, the selection of joists must be justified by calculation. Unlike structural steel (for which calculations are generally based on limit state principles), the structural use of timber is still usually determined by permissible stress principles, and these are set out in the BS 5268 series.

The new floor may also need to be capable of performing the following two structural functions:

- *Tying*: the new floor may be required to tie-in one or more roof slopes either directly, by connecting existing rafters to the new joists, or via intermediate elements linking the rafters to the new floor structure. This applies particularly where ceiling joists are removed below the new floor to provide increased headroom. Effective tying eliminates outward thrust generated by the roof slope, or slopes. An inadequately restrained roof slope will tend to spread the walls that support it, but as with many other forms of structural failure, it is not always immediately apparent and may only manifest itself over a number of years.
- *Lateral restraint*: the new floor *may* be required to provide lateral restraint for the walls of the building. In the case of a conversion that creates a new storey in an existing two-storey house, it is generally assumed that the original roof structure, with appropriately fixed wall plates, rafters and ceiling joists, provides adequate lateral restraint, at least in the direction of the existing ceiling joists. However, as with tying (above), any alteration of the ceiling joist/rafter foot/wall plate configuration may reduce the ability of a wall to resist movement at right angles to its plane. In addition, the floor will be required to provide restraint where a new gable wall is raised.

In addition to its load-bearing functions, it is of equal importance that the floor (including the ceiling beneath it) be capable of providing the following:

- *Fire resistance*: the new floor in the loft conversion of a single family dwelling, and other intermediate floors, must provide appropriate fire resistance. In single family dwellings, a 30-minute standard is required.
- *Sound resistance*: the floor of the conversion must provide resistance to the passage of sound.
- *Moisture resistance*: in the case of bathrooms, kitchens and utility rooms, any board used as flooring should be moisture resistant.

ELEMENTS OF LOFT CONVERSION FLOOR DESIGN

The availability of headroom is an important factor in most loft conversions. Because planning considerations mean that raising the ridge height of a roof is usually not feasible, the overall thickness of the floor structure must be carefully considered in order to create reasonable room heights. A number of strategies are adopted to minimise the impact of floor thickness on available headroom. These include:

■ *Minimising joist span*. Deeper joist sections are required on extended spans, and it is generally prudent to limit clear spans to about 4 m. The use of intermediate structural support, which could be provided by beams or perhaps suitable internal walls, may be considered. Note that the maximum span of any floor supported by a wall is 6 m (Approved Document A *Structure* 2C23).

■ *Position of new floor joists*. To optimise available headroom, new floor joists are generally run parallel to the existing ceiling joists, which are retained. This is usually achieved either by partially suspending the new joists from support beams (underslinging) or by packing them up off existing wall plates. Note that the spacing of the new floor joists will be influenced by the position of the existing ceiling joists – see note on joist spacing below. In traditionally built houses (certainly pre-1950) with cut roofs, ceiling joist/rafter spacing is not necessarily equal.

■ *Dropping lower floor ceilings*. This approach may be used to gain headroom where clearances are particularly limited: the lower floor ceiling and supporting ceiling joists are removed and the new conversion floor structure inserted. This approach has the disadvantage of reducing ceiling height in the rooms immediately below the conversion, and causes considerable disruption to the occupants of the building. It also requires a thorough assessment of the function of the existing pitched roof structure: note that ceiling joists perform an essential tying function in most roofs and they should not be removed unless an alternative means of tying is provided. Alternative methods might include full restraint of rafters by a suitably designed ridge beam and purlins, and the introduction of new low-level ties.

Room height in the conversion (headroom)

As noted in Chapter 5, there is no longer a stated minimum floor-to-ceiling height for rooms but it is generally accepted that a minimum of 2.3 m be maintained over the main portion of the conversion. Rooms with ceilings below this height are somewhat oppressive.

Proposals for reduced floor-to-ceiling heights within rooms should be carefully considered in relation to any downward projection from a ceiling, for example, a downstand beam. Where this is the case, it would be reasonable to consider guidance included in Approved Document K *Protection from falling, collision and impact* and to ensure that the clearance between any such projection and floor level exceeds 2 m. Headroom on stairs and landings is considered at the end of the chapter.

METHODS OF SUPPORT FOR FLOORS IN LOFT CONVERSIONS

Floor joists in a conversion may be supported in two ways:

(1) *Indirectly*: new floor joists are supported by beams which transfer the load to walls or other suitable load-bearing structure. Joists are either fixed into beam

webbing or are supported by hangers nailed to timber bearers that are bolted to the beam.

(2) *Directly*: new floor joists are supported directly by existing walls. Joists may be fixed to the tops of walls (on wall plates) or to the sides of walls with masonry joist hangers.

Depending on circumstances, a combination of direct and indirect methods may be used.

(1) Beam-supported floors (indirect)

This is the most commonly used method of supporting the new floor in a loft conversion. In a typical example, bearings are provided and steel beams fixed between supporting walls (note the requirements of the Party Wall Act – see Chapter 1). To reduce the overall thickness of the floor structure, beams are set at right angles to existing ceiling joists to allow new floor joists to 'float' between them (underslinging). It is usually necessary to remove ceiling binders where this is the case (see *Binders* below). As with floor joists, it is practice to leave at least 25 mm clearance between the supporting steel beams and the ceiling joists to prevent damage caused by deflection to ceilings in rooms directly beneath.

There are various methods for fixing timber joists to steel. The approach adopted will depend on the position of new floor joists relative to the beam. There are three principal approaches:

- Underslung floor joists
- Floor joists into beam webbing
- Additional support for existing ceiling joists

Underslung floor joists

Underslinging is used where it is necessary to partially suspend joists from a beam (Fig. 7.1). This would apply, for example, where the new joists must 'float' between existing ceiling joists in order to gain headroom in the conversion. This approach can only be applied to support the floor at the perimeter of the conversion: because the beam is set above the level of the floor, it would clearly not normally be possible to provide *intermediate* support to the floor structure in this way.

The specification of appropriate joist hangers is of the utmost importance in this application, and hangers specifically designed to undersling joists are produced by most of the major fixings manufacturers. These are often generically described as 'long leg' or 'extended leg' hangers, and a number of proprietary versions are available. All work similarly to conventional timber-to-timber hangers, the chief difference being the length of the legs, which are designed to wrap over the carrying member. Note that the safety and stability of floors supported in this way are almost entirely dependent on correct nailing and wrap-over, and it is therefore essential that the manufacturers' nailing schedules be adhered to.

Long-leg hangers cannot be fixed directly to steel and it is therefore necessary to fix timber carrying members into the web and a bearing plate to the top flange of

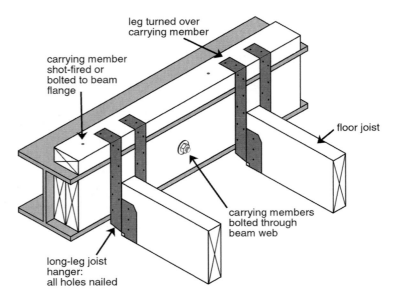

leg turned over
carrying member

carrying member
shot-fired or
bolted to beam
flange

floor joist

carrying members
bolted through
beam web

long-leg joist
hanger:
all holes nailed

Fig. 7.1   Underslung floor joists.

the beam to permit nailing. Methods of fixing bearers and packing timber to steel beams include bolting, shot-firing and the use of self-drilling screws.

The amount of joist downstand that may be achieved relative to the supporting beam depends on the design of the hanger: manufacturers' technical guidance must be followed. Typical maximum downstand for proprietary hangers is generally between 75 and 100 mm.

However, a careful assessment of relative downstand should be made where existing ceiling joists are to be retained in proximity to new floor joists: in some cases the leg flange at the base of the joist hanger will foul the adjacent ceiling joist.

Floor joists into beam webbing

This configuration is generally used where the beam must be incorporated within the structural thickness of the floor, for example, where the floor, and perhaps the lower ceiling surface as well, must continue uninterrupted above and below the beam (Fig. 7.2). The cut end of the joist is accommodated within the web of the beam (often a universal column in this application) and supported on its bottom flange. Timber blocking should be fixed between joist ends in the web to limit lateral and torsional movement.

It is generally necessary for the joist to be scribed into the webbing with this arrangement. For example, where the joist is required to support floor and ceiling materials carried over the beam, both the top and bottom surfaces of the joist should be proud of the flanges (as indicated in Fig. 7.2) to accommodate timber shrinkage that might otherwise draw finishes into direct contact with the beam. It would normally be necessary to provide calculations to justify notching joists at their bearing points.

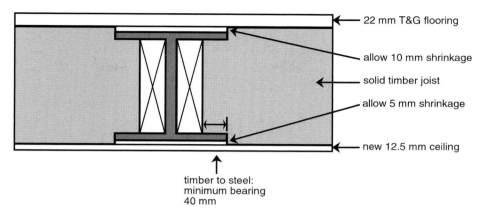

Fig. 7.2 Floor joists into beam webbing.

This approach may be used to support new floor joists above existing ceiling joists within the roof space, although obviously with a loss of headroom. It is also used where the ceiling of the lower floor is 'dropped' to gain headroom.

Additional support for existing ceiling joists

Although this approach depends on support provided by beams, it is quite different in principle from the approaches described above: beams are fixed at appropriate intervals *beneath* and at right angles to the ceiling joists to reduce their effective span as illustrated in Fig. 7.3. Thus supported, the existing ceiling joists

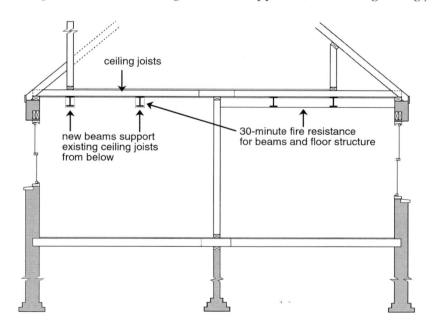

Fig. 7.3 Additional support for existing ceiling joists.

are then used as floor joists in the conversion. The advantage with this approach is that it allows headroom to be maintained in the conversion and considerably reduces the amount of timber required to construct the floor because new floor joists are not required.

This system is seldom adopted for a number of reasons, not least the disruption it causes to the occupants of the building. The following practical points would require consideration:

■ *Existing ceiling joists and supports*: the adequacy of wall plates and any internal load-bearing walls must be assessed and the sectional dimensions, span and spacing of ceiling joists noted. In most cases, these joists will not be grade marked, and it will be necessary to make a judgement on their strength class based on visual assessment.

■ *Buildability*: the practical constraints of handling beams should be considered particularly in relation to obstacles such as partition walls through which they must pass. In order to make steel beams manoeuvrable, it would generally be necessary to split them into sections with spliced connections designed by a structural engineer (see Chapter 6).

■ *Relative beam position*: beams are installed at ceiling level in the floor immediately below the conversion. Ceiling materials, such as plasterboard or lath and plaster, must be cut away to allow direct contact between the beam and ceiling joists. The position of beams is influenced by the requirement to limit the span of the joists and by the presence of obstructions such as chimney breasts and flues: it is not permissible to fix structural elements, such as beams and joists, directly to chimney breasts or flues. In many cases, the relatively close spacing of the supporting beams will limit space for staircase access to the conversion and it will be necessary to introduce additional beams to trim a suitable floor opening.

■ *Effect on lower floor rooms*: beams installed at ceiling height on the lower floor will clearly reduce floor to ceiling height for rooms and circulation spaces at the lower level.

■ *Fire resistance of beams*: because floor beams must be protected from fire, it will be necessary either to clad them with an appropriate material or to construct an entirely new ceiling beneath them offering 30-minute fire resistance (see *Floor fire resistance* below). The use of intumescent paint may also be considered. The aesthetic impact of exposed steel beams in habitable rooms is also a factor to consider.

(2) Wall-supported floors (direct)

Even the structural design of loft conversions is subject to fashion and there is a tendency, certainly in routine conversions, to specify the use of steel beams without even investigating the possibility of supporting joists directly on existing elements of structure.

Most traditionally constructed dwellings and a high proportion of contemporary dwellings include structural elements that may have the potential to provide

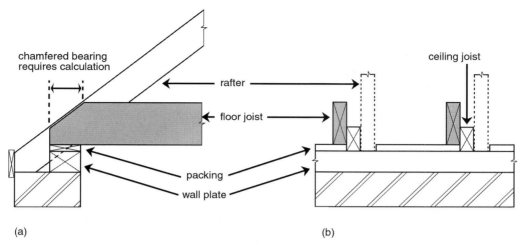

Fig. 7.4   Wall-supported floor joists.

support for a new floor directly (Fig. 7.4a and b). These include existing wall plates set on opposing external load-bearing walls and, particularly in pre-war dwellings, a central spine wall that in many cases is configured for a proportion of roof loading.

A thorough structural assessment must be carried out before undertaking the design of such a conversion. Particular attention should be paid to the following:

- *External walls*: wall plates must be adequately fixed to load-bearing walls, be properly bedded in mortar and be in good condition. The position of windows, doors and other openings relative to wall plates – and the general condition of the wall – must be considered.
- *Internal walls*: the spine wall, or other internal walls proposed for use, must be of load-bearing construction. In addition, it will be necessary to check that such walls are adequately restrained, are in true vertical continuity between floors and have adequate footings.
- *Rafter/wall plate relationship*: if it is necessary to notch or chamfer the top of the joists to accommodate them beneath the pitch of the roof, this will generally need to be justified by calculation (Fig. 7.4a).
- *Lintels*: where it is proposed to support joists on a wall plate or in masonry above a lower floor window, it would generally be necessary to prove the structural adequacy of the lintel, perhaps by exposing it, and to replace it if necessary. A less invasive approach would be to trim around the window opening (Fig. 7.5).
- *Bays*: bay window structures may have limited load-bearing capacity and should be treated with caution. As with lintels (above), trimming could be considered as an alternative means of support.
- *Buildability*: in addition to the structural factors outlined above, the potential difficulties of providing adequate fixings for joists in often confined eaves spaces should be considered (Fig. 7.4b).

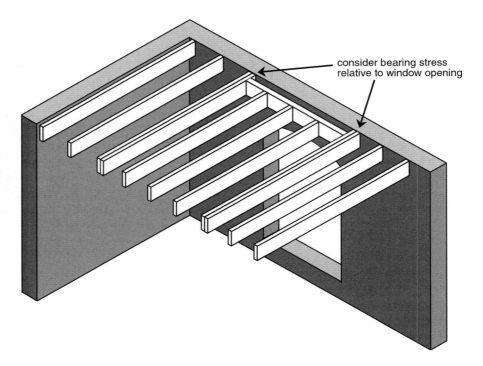

consider bearing stress
relative to window opening

Fig. 7.5   Wall-supported floor joists: trimming.

- *Span*: in order to limit the load on walls, guidance provided in Approved Document A *Structure* limits the span for any floor supported by a wall to 6 m, with span measured from the centre of the bearing.
- *Structural clearance*: floor joists should be set on packing to provide 25 mm clearance from the existing ceiling structure.

FLOOR JOIST SELECTION

The floor joists used in domestic conversions are generally produced from solid timber that is graded according to strength. This is often generically described as softwood carcassing timber. In domestic conversions, timber of strength classes C16 (similar in strength to the former SC3) or C24 (similar to the former SC4 and stronger than C16) is generally used. It is also possible to use timber I-joists in this application.

Unlike beams, there is no requirement to provide structural engineering calculations to justify the size of solid timber floor joists (unless additional loads must be considered – see beginning of chapter) and sizing may be determined using timber tables. In all cases, however, it is important that timber of the appropriate strength class is specified – and used – during construction. Graded timber is marked to indicate conformity with its particular strength class and it is important that any timber used in structural applications displays such a mark in order

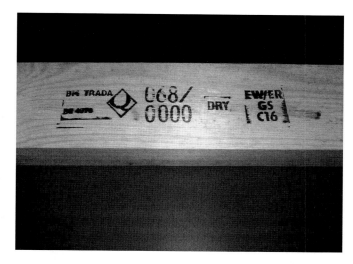

Fig. 7.6   BM TRADA Certification timber grade mark. Courtesy TRADA Technology Ltd.

to demonstrate compliance. A BM TRADA Certification grade mark is illustrated in Fig. 7.6.

Span tables for joists and other timber elements were formerly provided in Appendix A of Approved Document A *Structure* (1992). However, the adoption of the current timber strength classes (e.g. C16, C24) led to the need for revision of earlier span tables. The new tables – *Span tables for solid timber members in floors, ceilings and roofs for dwellings* – are published by TRADA Technology. These tables are referred to but not included in the current 2004 edition of Approved Document A.

It should be noted that, in most cases, spans for comparable sections are marginally greater in TRADA's revised tables. These tables are intended for use in single occupancy dwellings of up to three storeys. They are not intended for use with trussed rafter roofs. As with the earlier span tables published in Approved Document A, concentrated loading (for example, loading introduced by trimmers and partitions) is not taken into account. However, certain common loads, such as those caused by baths or lightweight partitions, may be accommodated provided that rules on joist duplication and span modification are observed (see baths and partitions below). Note that joists may also be specified on the basis of calculation, and procedures for this are set out in BS 5268.

Joist spacing

In buildings constructed until about 1970, spacing generally conforms to imperial dimensions, with typical centre-to-centre measurements of 14" (approximately 356 mm), 16" (approximately 406 mm) or 18" (approximately 457 mm). As noted earlier, however, the spacing of ceiling joists and rafters in traditional cut roofs is not necessarily regular; indeed, it was often a somewhat haphazard affair with spacing varying considerably.

Table 7.1   Commonly available joist sections*.

47 × 150 mm	75 × 150 mm
47 × 175 mm	75 × 200 mm
47 × 200 mm	75 × 225 mm
47 × 225 mm	

* The UK imports a high proportion of its carcassing timber; the sections above are typical of European-sourced stock at dockside in the UK in July 2005.

Timber supplies

Whatever the technical basis for joist selection, it should be noted that the stock availability of given section sizes, and lengths, is also a determining factor. For example, timber with a breadth of 47 mm is the most widely used and represents a high proportion of timber imported to the UK. By contrast, 50 mm timber is rather less common. Any specification that requires a special order is likely to add to costs and may take longer to obtain.

Span tables recognise and provide data for 40 different breadth/depth configurations for joists representing customary sizes. However, some of these sections are more readily available than others (Table 7.1).

Machined (regularised) joist sections

The section sizes quoted by wholesalers represent the nominal dimensions of timber as it rolls off the producer's sawmill. However, timber should be machined (regularised) on width (depth) to provide an even surface for floors and ceilings. TRADA's *Span tables* (and the earlier Approved Document span tables) take this additional processing into account: for example, 47 × 150 mm is not referenced as a joist section size, but 47 × 145 mm is. Note the conventions for expressing the major and minor dimensions of timber sections, and particularly the use of the word 'width' in this context (Fig. 7.7).

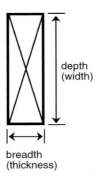

depth
(width)

breadth
(thickness)

Fig. 7.7   Joist dimensions.

Table 7.2 Floor joists (C16) – permissible clear spans.

Joist size (mm)	Joist spacing (mm)	Clear span (m)
47 × 97	400	1.67
47 × 147	400	2.70
47 × 170	400	3.12
47 × 195	400	3.54
47 × 220	400	3.95
75 × 147	400	3.22
75 × 170	400	3.71
75 × 220	400	4.74

Based on Approved Document A (1992). TRADA's span tables (2005) give slightly greater spans in all but one example.
Imposed load not exceeding 1.5 kN/m^2
Dead load 0.50 to 1.25 kN/m^2 (excluding self weight of joist)

Table 7.3 Floor joists (C24) – permissible clear spans.

Joist size (mm)	Joist spacing (mm)	Clear span (m)
47 × 97	400	1.80
47 × 147	400	2.92
47 × 170	400	3.31
47 × 195	400	3.79
47 × 220	400	4.26
75 × 147	400	3.34
75 × 170	400	3.86
75 × 220	400	4.88

Based on Approved Document A (1992). TRADA's span tables (2005) give slightly greater span in all but one example.
Imposed load not exceeding 1.5 kN/m^2
Dead load 0.50 to 1.25 kN/m^2 (excluding self weight of joist)

Loading conventions in tables

The data in TRADA's *Span tables for solid timber member in floors, ceilings and roofs for dwellings* is arranged in a similar manner to that in the appendix of old Approved Document A (Tables 7.2, 7.3). A number of factors are considered, including:

- *Strength class*: C24 is stronger than C16 and permissible spans are therefore greater. For example, spaced at 400 mm centres, 47 × 220 mm joists subject to a dead load not exceeding 1.25 kN/m^2 in strength class C16 will safely span 3.99 m. A joist of the same sectional dimensions in C24 would span an additional 420 mm.
- *Imposed load*: the load is assumed to be no greater than 1.5 kN/m^2. This figure represents the loading generated by people and furniture. It does not include structural loading.
- *Dead loads*: three categories of dead loads are considered. These are:
  (1) not more than 0.25 kN/m^2
  (2) more than 0.25 but not more than 0.50 kN/m^2
  (3) more than 0.50 but not more than 1.25 kN/m^2

Note that the dead load of materials needed to provide fire and sound resistance means it is very unlikely that a new floor in a loft conversion would fall into the first category (i.e. not more than 0.25 kN/m²). Determining which of the remaining two categories the floor would come under requires an assessment of the anticipated dead loads.

Sometimes a common overall value of 0.60 kN/m² is assumed when lofts are converted, and joists would therefore be selected from the third group (i.e. more than 0.50 but not more than 1.25 kN/m²). Note that this does not include any additional loads from roofs or walls.

Bath support

In the case of a standard bath, it is generally not necessary to justify the loading by calculation when using *Span tables*, provided that joists supporting it are doubled. Note that where joists are duplicated, they must be fixed together (Fig. 7.8). The use of 51 mm diameter double-sided round toothed plate connectors (double-sided connectors) and M12 grade 4.6 bolts at 600 mm centres may be considered.

Partition wall support

Certain lightweight partitions may be erected without the need for structural engineering calculations provided that the guidance provided in *Span tables* is followed. Such partitions must weigh no more than 0.8 kN per linear metre and must not be load-bearing.

In the case of partitions running parallel to joists, up to two additional joists must be provided dependent on spacing. These should be connected to each other as outlined in bath support (above).

In the case of partitions running at right angles to the joists, maximum spans must be reduced by 10%.

Partitions should be fixed through the flooring material into the joists below. Where partitions are load-bearing or exceed 0.8 kN/m, joist size must be determined by calculation. Note that a sound-resisting partition wall comprising

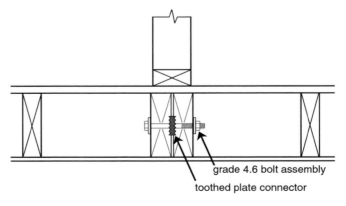

Fig. 7.8   Double joists.

100 mm studwork with plasterboard/skim finish and mineral wool infill is likely to have a unit load of more than 0.35 kN/m². At a height of 2.3 m, such a wall is likely to marginally exceed 0.8 kN/m.

Roof load

Note that where a timber floor must carry any portion of roof loading (see Fig. 9.6), it is not possible to use span tables to select joists and calculations must be carried out to BS 5268.

Minimum bearing for floor joists

Guidance provided in *Span tables* indicates that a minimum bearing area for joists of 40 mm should be provided at supports (Fig. 7.9). Note that this is slightly more than the figure included in the earlier (1992) Approved Document A *Structure* which was 35 mm. It is emphasised that 40 mm is a minimum and, where joists are to be fixed, by skew nailing to a wall plate for example, it would clearly be prudent to allow considerably more than this to provide an adequate fixing. It would normally be necessary to justify by calculation any proposal for a bearing less than 40 mm. In addition, it should be noted that an increased bearing is required if the floor is to provide lateral support for walls (see *Lateral support by floors* below).

Holes and notches in joists

It is usually necessary to cut into or drill through joists in order to accommodate cables and pipes that run at right angles to the direction of the joists. Note that the rules for the safe positioning of holes and notches are quite different from each other:

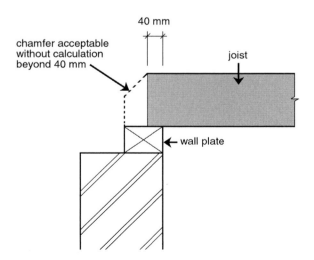

Fig. 7.9   Minimum bearing.

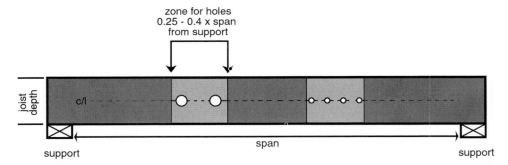

Fig. 7.10a   Hole positions (floor and ceiling joists).

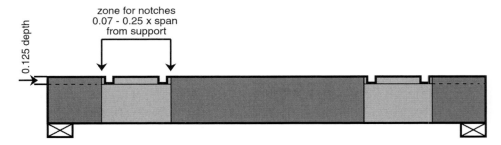

Fig. 7.10b   Notch positions (floor and ceiling joists).

- *Holes*: should be drilled at the centreline (neutral axis) of the joist (Fig. 7.10a). The diameter of holes should not be greater than 0.25 times the depth of the joist and they should not be less than three diameters apart, measured from centre to centre. The safe zone for holes starts 0.25 of span length away from the support and ends 0.4 times the span length away.
- *Notches*: the safe zone for notches lies between 0.07 and 0.25 times the span length away from the support (Fig. 7.10b). Notches should not be deeper than 0.125 times the depth of the joist.

Note that notches or holes should generally not be cut in rafters, purlins or binders unless justified by structural engineering calculation.

BINDERS

In traditional 'cut' roof construction, ceiling joists were generally fixed by nailing at wall plates at each end of their span and also to the feet of the rafters for which they provide a tie in the majority of cases. However, in order to reduce the sectional dimensions of the timber needed to support the ceiling, intermediate support is often provided by binders (sometimes called ceiling bearers, runners or hanging beams) that are fixed above, and run at right angles to, the ceiling joists (Fig. 7.11). Binders are also found in TDA trussed roofs. Note that the binders

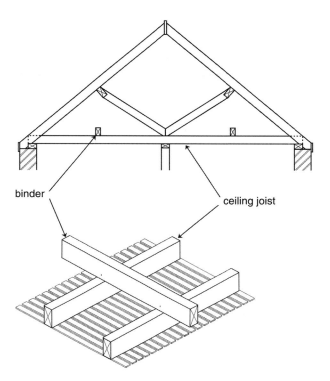

Fig. 7.11 Binders.

used in modern trussed rafter roofs are generally configured to resist lateral loading.

Binders in cut roofs are supported at each end and sometimes at intermediate points on span, and act as beams as far as the ceiling structure is concerned. Two additional versions of the conventional binder are sometimes encountered. One takes the form of a section of timber fixed to the top of the ceiling joists broad side down. In this configuration, the 'binder' may not be supported at its ends. In this sense it is not a true binder, but such an element does provide a degree of load sharing across the ceiling joists.

In another configuration, the binder may provide support not only for the ceiling joists but also for purlin struts. A binder of this sort (sometimes called a strutting beam) is subject to both roof and ceiling loads (see Fig. 9.4c).

In order to gain headroom in the conversion, it is usually necessary to remove binders to allow new floor joists to be positioned between the existing ceiling joists. When binders are removed, a new way of providing intermediate support for the existing ceiling must be found. Where spacing allows, this may be achieved by positioning the new floor joists next to the ceiling joists: support for the ceiling joists may then be provided by strapping, screwing or nailing them directly to the new floor joists.

Strapping – using lengths of perforated galvanised steel band – is generally preferable to nailing or screwing. The strapped connection offers a degree of

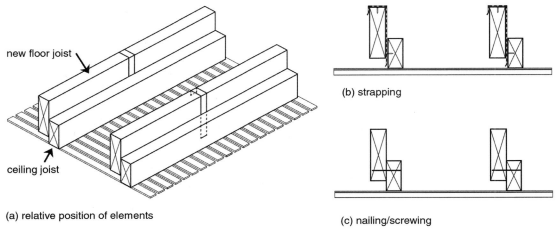

(a) relative position of elements

(b) strapping

(c) nailing/screwing

Fig. 7.12   Binder replacement.

flexibility, and therefore any small deflection of the new floor joists is less likely to result in damage to ceiling surfaces (Figs 7.12a and b). An alternative is to nail or screw through the existing ceiling joists to the new floor joists (Fig. 7.12c) or to provide noggings, although these both have the obvious disadvantage of creating rigid connections. Note that the act of nailing through ceiling joists may cause a degree of damage to a relatively fragile ceiling surface.

Where straps or other connections are to replace binders, it would be reasonable to provide the new supports at same point on-span as the original binder on the grounds that this arrangement would be no worse than the original. In all cases, it will be necessary to provide temporary support for the ceiling whilst binders are removed.

New floor joist to existing ceiling clearance

It is practice to maintain a clearance of 25 mm between the underside of new floor joists and the existing lower floor ceiling. The primary reason for doing so is that it reduces the risk of damage to the ceiling caused by joist deflection.

In terms of buildability, there is also a strong argument for leaving a reasonable clearance between old and new floor elements. This applies particularly in older structures where it is unlikely that the ceiling will be level across the width of the building, and a degree of relative concavity is likely between supports.

STRUTTING

Strutting, sometimes described as blocking or nogging, is fixed between floor joists (and flat roof joists) to counteract the timber's natural tendency to twist as it dries out. It is also effective in enhancing load sharing between joists. In addition, strutting helps to reduce vibration in floors.

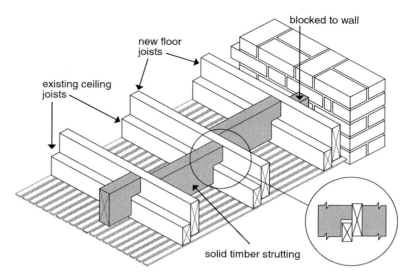

Fig. 7.13   Solid timber strutting.

Strutting should be provided at the supported ends of joist runs (for example, between joist ends within beam webbing). However, intermediate (mid-span) strutting is not generally required for spans less than 2.5 m. For joists with a span of more than 2.5 m, strutting is provided mid-span. Where the span exceeds 4.5 m, two rows of strutting should be provided with one at each third point along the span. Note that the breadth-to-depth ratio of joists may also influence the location of strutting. Strutting should be blocked solidly to perimeter walls.

Solid timber strutting, timber herringbone strutting and proprietary strutting devices may all be used in this application. However, in order to fix diagonal bracing such as herringbone and proprietary strutting systems, access is required from below as well as from above the joist. Unless the lower floor ceiling and its supporting joists have been removed as part of the conversion, it is not possible to use these approaches, and solid strutting should be used instead.

Solid struts, a minimum of 38 mm in breadth, should be less than the full depth of the joist to minimise the risk of disturbing floor or ceiling surfaces. Guidance is for a minimum of 0.75 depth. In a loft conversion where existing ceiling joists project into the void between the new floor joists, it is not always possible to fix strutting of minimum depth across the full width between the joists, and it is therefore necessary to notch the struts to accommodate the ceiling joist projection. Where this is necessary, it would be reasonable to provide a clearance of at least 25 mm between the strut notch and the ceiling joist to accommodate deflection (Fig. 7.13).

TRIMMING

Trimming members are introduced to provide support where joist runs are interrupted by projections or openings in the floor structure. In a loft conversion, there are typically two sets of circumstances where trimming is necessary:

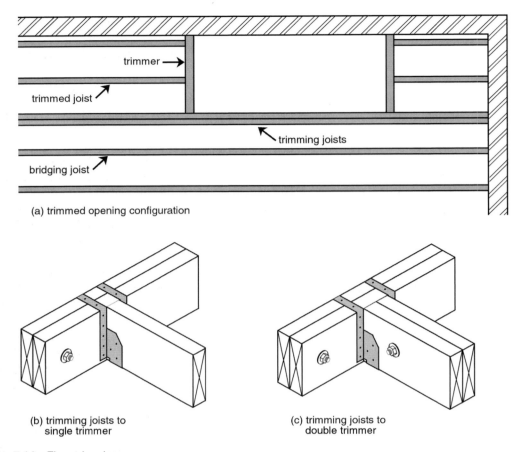

(a) trimmed opening configuration

(b) trimming joists to single trimmer

(c) trimming joists to double trimmer

Fig. 7.14   Floor trimming.

■ Around the stairwell opening to the new floor
■ Around a chimney breast

A third set of circumstances under which floor trimming may be employed occurs where new floor joists must be supported by an existing masonry wall that contains an opening with a lintel of unknown strength. It would clearly not be acceptable to introduce a new loading directly above such an opening and it is therefore practice either to fix a lintel of known strength or to trim around the opening (Fig. 7.5).

Where primary support for the new floor structure is to be provided by steel beams, it is becoming increasingly common to use secondary steel beams as trimming members for stairwell openings and projections such as chimney breasts. In the case of chimney breasts, the secondary steel beam may perform a dual role, providing both a trimming joist for the floor and also support for a new ridge beam via a timber post (see also Fig. 6.1). However, the conventional approach using timber trimming is still widely used in floor framing (Fig. 7.14). There are typically four elements in a trimmed floor:

(1) *Bridging joists*: bridging or common joists run between their natural supports without interruption.
(2) *Trimmed joists*: bridging joists that are cut short to accommodate the opening or projection.
(3) *Trimmer joist*: a joist that runs at right angles to, and provides support for, the cut ends of the trimmed joists.
(4) *Trimming joists*: trimming joists run parallel to the bridging joists and provide support for the trimmer.

In effect, the trimming and trimmer joists behave as beams. Because they are subjected to high point loads it is necessary that they be made stronger than the trimmed joists they support. The requirement to provide a regular surface for ceilings and floors means that it is generally not practical to provide deeper joists and it is therefore practice to provide trimming members of increased breadth. Note that trimming joists must be fixed together with bolts and toothed plate connectors as illustrated in Fig. 7.8.

The traditional approach to sizing trimming and trimmer joists supporting no more than six trimmed joists was to make them 25 mm thicker than the bridging joists. However, under current practice it would be unwise to proceed on this basis and trimming members should be determined by calculation.

Joints between members within the trimmed structure are now usually made using metal joist hangers or fixing plates (Figs 7.14b and c). Traditional joints, such as tusk tenon and blind tenon joints, are relatively time-consuming to produce and their effectiveness is only achieved through craftsmanship of a high order.

In designing the conversion, it is clearly desirable to limit the extent of trimming as far as possible, particularly where large openings in the floor are required. For example, ensuring a parallel relationship between a new stairwell and floor joists in a conversion will considerably reduce the need for trimming.

Note that it is also necessary to trim floor structures around chimney breasts (Fig 7.15). Guidance on clearances between combustible materials (such as timber joists) and flues is contained in Approved Document J. Note that there is a presumption against making any structural connections to chimney breasts.

LATERAL SUPPORT BY FLOORS

Guidance in Approved Document A *Structure* (section 2C) is that floors above ground level should be fixed to walls (Table 7.4). This allows the floor to act as a horizontal brace. Lateral support of walls by floors is effective in restricting the movement of external walls. Internal load-bearing walls also require lateral support. Note that a new roof may also provide lateral support for walls.

Formalised guidance on lateral support of walls by floors and roofs is, in historical terms, a relatively recent development. Traditional modes of construction meant that walls often had a reasonable degree of intrinsic stability with floor and ceiling joists fixed to walls providing lateral support in a longitudinal direction.

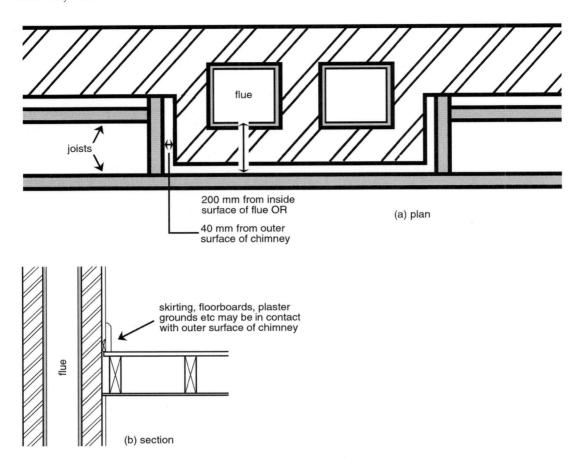

Fig. 7.15 Combustible materials: proximity to flue/chimney.

Table 7.4 Lateral support for walls provided by floors.

Wall type	Wall length	Lateral support required
Solid or cavity: external wall, compartment wall or separating wall	Greater than 3 m	Floor lateral support by every floor forming a junction with the supported wall
Internal load-bearing wall (not being a compartment or separating wall)	Any length	Floor or roof lateral support at the top of each storey

However, designed connections between floor joists and walls *parallel* to them (i.e. generally along flank walls) were not routinely provided.

Most loft conversions are, of course, carried out in traditionally constructed buildings. Where elements already providing lateral support for existing walls are removed or altered, it is necessary that restraint be provided in some other way. This may occur, for example, if ceiling joists are removed (but note also their independent tying function described earlier).

The position is less clear as far as previously *unrestrained* existing external or compartment walls are concerned. Examples of unrestrained walls may be found in the case of flank walls of detached, semi-detached and end-terrace dwellings. It would clearly be good practice to provide connections between the new floor and such walls, but this is sometimes overlooked.

However, where the conversion involves the construction of a new wall (as in a hip-to-gable conversion) or the extension of an existing one (for example, where a gable is extended to form a flank gable), the lateral restraint of those walls by the new floor must be considered. Fig. 7.16 illustrates methods of providing lateral restraint for cavity and solid masonry walls.

Where it is necessary for a new floor to provide lateral support, the guidance set out in Approved Document A *Structure* may be followed. However, Approved Document A lists a number of conditions where tension (restraint) straps need not be provided, including:

■ in the longitudinal direction of joists in houses of not more than two storeys, if the joists are at not more than 1.2 m centres and have at least 90 mm bearing on the supported walls or 75 mm bearing on a timber wall plate at each end; and

■ in the longitudinal direction of joists in houses of not more than two storeys, if the joists are carried on the supported wall by joist hangers in accordance with BS EN 845-1 of the restraint type described in BS 5628: Part 1 and are incorporated at not more than 2 m centres; and

■ where floors are at or about the same level on each side of a supported wall, and contact between the floors and wall is either continuous or at intervals not exceeding 2 m. Where contact is intermittent, the points of contact should be in line or nearly in line on plan.

FLOOR FIRE RESISTANCE

In order to protect occupants and structure from the effects of fire, guidance set out in Approved Document B *Fire safety* is that the new floor in a loft conversion should have a specified standard of fire resistance.

Note that the fire resistance of *all* intermediate floors in the dwelling, and elements of structure that support them, must be assessed and upgraded if necessary when a loft is converted. Normally, full 30-minute fire resistance is required for floors, but under certain circumstances (outlined in Chapter 4), a modified 30-minute standard is permissible at first floor level.

Without testing a given floor structure, it is impossible to ascertain precisely what level of fire resistance it would offer. However, a number of assumptions are made about existing floor structures. For example, it is assumed that a plaster on wood-lath ceiling of between 15 and 22 mm thickness *might* provide 20 minutes' fire resistance if it is in good condition (i.e. plaster nibs intact and forming an unbroken bond with the laths). A thicker ceiling would not necessarily perform any better and, indeed, might fail more rapidly. It is also assumed that plain-edge floorboards, or badly fitting tongued and grooved boards, contribute little to overall fire resistance (Fig. 7.17).

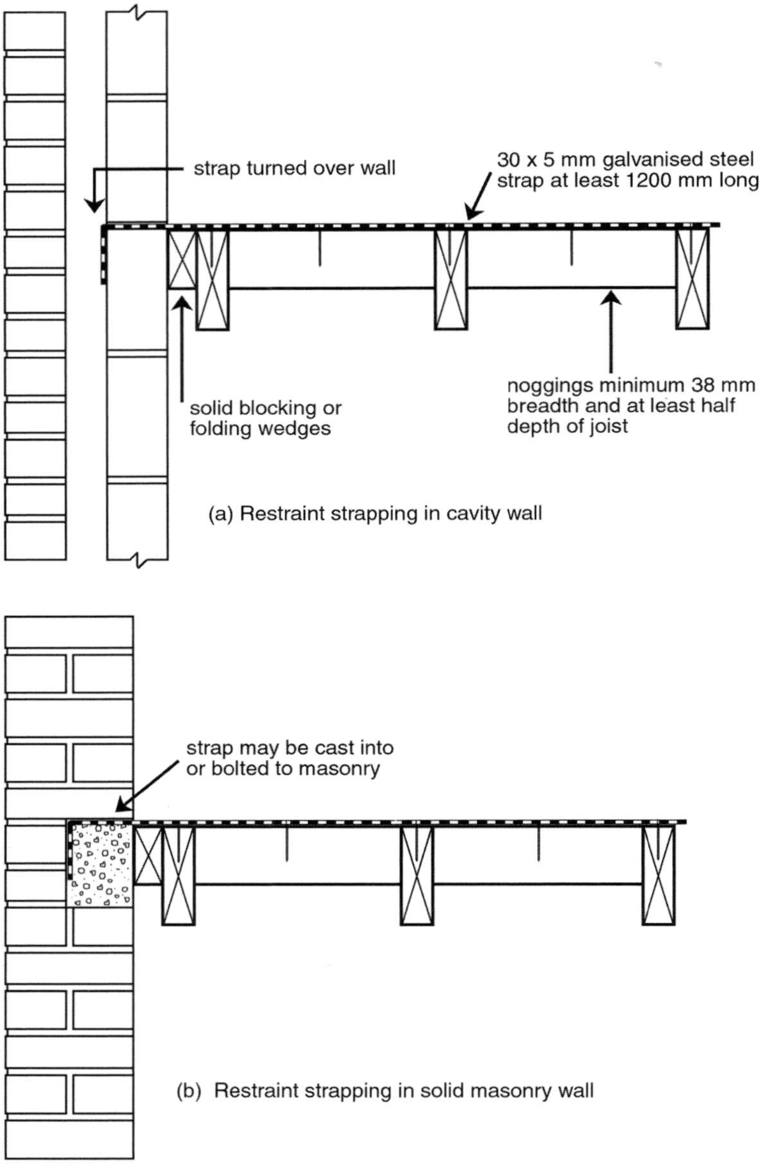

strap turned over wall

30 x 5 mm galvanised steel
strap at least 1200 mm long

solid blocking or
folding wedges

noggings minimum 38 mm
breadth and at least half
depth of joist

(a) Restraint strapping in cavity wall

strap may be cast into
or bolted to masonry

(b) Restraint strapping in solid masonry wall

Fig. 7.16   Lateral support of walls by floors.

Given that there is a general reluctance to modify ceilings for practical and sometimes aesthetic reasons, remedial measures are applied from above rather than below wherever possible. These measures might include the introduction of mineral wool pugging between floor joists. This contributes to fire resistance provided it is brought into close contact with the joists. In addition to this, overlaying plain-edge boards or badly fitting T&G flooring with either 3.2 mm hardboard or 4 mm ply will also enhance fire resistance.

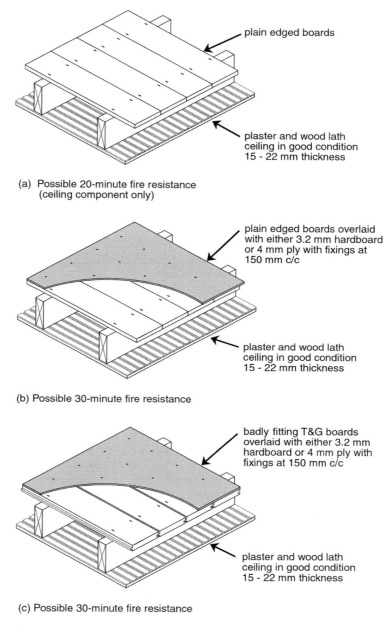

(a) Possible 20-minute fire resistance
(ceiling component only)

plain edged boards

plaster and wood lath
ceiling in good condition
15 - 22 mm thickness

plain edged boards overlaid
with either 3.2 mm hardboard
or 4 mm ply with fixings at
150 mm c/c

plaster and wood lath
ceiling in good condition
15 - 22 mm thickness

(b) Possible 30-minute fire resistance

badly fitting T&G boards
overlaid with either 3.2 mm
hardboard or 4 mm ply with
fixings at 150 mm c/c

plaster and wood lath
ceiling in good condition
15 - 22 mm thickness

(c) Possible 30-minute fire resistance

Fig. 7.17   Floor fire resistance: existing floors.

In all cases, the ceiling must be imperforate in order for floor fire resistance requirements to be met. The installation of recessed lighting (ELV luminaires for example) is likely to compromise the integrity of the ceiling. In all cases where fire resistance is required, luminaires should be protected by appropriate fire hoods installed according to the manufacturer's instructions.

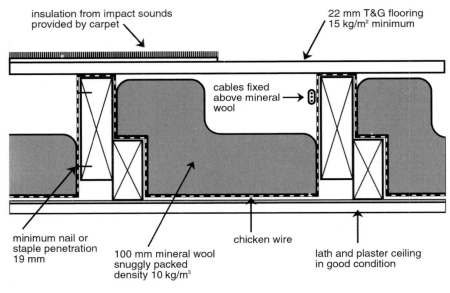

insulation from impact sounds
provided by carpet

22 mm T&G flooring
15 kg/m² minimum

cables fixed
above mineral →
wool

minimum nail or
staple penetration
19 mm

100 mm mineral wool
snuggly packed
density 10 kg/m³

chicken wire

lath and plaster ceiling
in good condition

NOTE: recessed light fittings
to be fitted with fire hoods

Fig. 7.18   Conversion floor: section.

Conversion floor (fire and sound resistance)

The new conversion floor must comply with guidance on fire resistance set out in Approved Document B *Fire safety*. It should also conform to guidance on sound resistance (Approved Document E *Resistance to the passage of sound*).

Fire and sound resistance are, to a certain extent, complementary functions. Ceiling finishes and mineral wool of appropriate specification possess both fire and sound resisting properties; flooring (i.e. boarding) also contributes to fire and sound resistance. Fig. 7.18 indicates a possible configuration. Note that there is currently no requirement to retrospectively provide sound insulation for existing floors elsewhere in the dwelling.

FLOOR MATERIALS AND FIXING

To allow access for first-fix plumbing and first-fix electrical work, and to facilitate inspection of the floor structure as part of the building control process, boarding is not generally immediately fixed in its final position after joists are installed. Instead, boards may be fixed selectively until work has advanced to a point where it is practical to permanently fix the flooring. Whilst fixing techniques vary depending on the type of flooring selected, there are three basic principles common to most boarding materials.

Conditioning

Most modern composite materials intended for flooring have a high degree of dimensional stability provided they remain dry. However, many types of board may benefit from being conditioned, or acclimatised, by being laid loose and flat in situ before being fixed. In the case of softwood timber boards, a degree of shrinkage must be anticipated, even with heavy cramping, in all but the most highly seasoned timber.

Staggered joints

Boards should be laid with joints staggered. Heading joints should be avoided because they reduce load sharing within the floor structure. In the case of exposed timber boards, rows of heading joints are also considered unsightly.

Moisture resistance

Note that in rooms where water may be spilled, such as bathrooms, utility rooms and kitchens, the guidance in Approved Document C *Site preparation and resistance to contaminants and moisture* (2004) is that flooring should be moisture resistant. Chipboard should conform to moisture resistance as specified in BS 7331:1990 or BS EN 312 Part 5:1997. These boards must be laid with identifying marks uppermost to demonstrate compliance. If softwood boards are to be used, they should be no less than 20 mm thick and from a durable species as set out in BRE Digest 429. Alternatively, softwood boards should be treated with a suitable preservative. Note that, to conform to guidance set out in Approved Document E *Resistance to the passage of sound*, flooring should have a minimum mass of $15 \, \text{kg/m}^2$.

Fixing

In order to prevent creaking, flooring should be fixed with either ring shank nails or screws. The joints of tongued and grooved panels should be glued. Note that conventional wire and cut nails have relatively poor resistance to withdrawal.

T&G floor panels

Flooring-grade tongue and groove chipboard panels are widely used, with a typical board size of 2440×600 mm (or 2400×600 mm) and a typical board thickness of 22 mm. The interlocking edges of these boards allow them to be laid with long edges running at 90 degrees to the joists. However, where long edges meet walls, 50×75 mm support noggings should be provided to prevent the boards from sagging. Short edges bear on joists. The boards are staggered to increase rigidity, and joints should be glued with waterproof PVA (polyvinyl acetate) adhesive to reduce the risk of creaking.

Traditional floorboards

Either T&G or plain-edged boards may be used for flooring. To conform to Approved Document E, these will need to have a minimum mass of 15 kg/m². In most cases, plain-edged boards would need to be overlaid with hardboard to conform to fire safety requirements. Note that sanded and sealed T&G floorboards, whilst attractive, are best restricted to ground floor use in the interests of impact noise reduction.

STAIRS

Guidance on stair access is provided in Approved Document K *Protection from falling, collision and impact*. The guidance is relatively straightforward and it includes a number of important concessions specifically for loft conversions. Additional information on stair specification can be found in the BS 585 and BS 5395 series of standards.

Headroom

Although there is no longer a minimum floor to ceiling height within rooms, Approved Document K (AD K) indicates that a minimum of 2 m headroom be maintained on stairs and landings (Fig. 7.19a). This should be considered with particular reference to any downward projecting elements (such as beams and purlins) which present an impact hazard.

In the case of loft conversions where it is not possible to achieve 2 m clearance, AD K indicates that reduced headroom is permissible with 1.9 m to the stair centreline and a minimum of 1.8 m on one side only (Fig. 7.19b). Downstand elements must be considered in this configuration as well.

Landings

Landings should be provided at the top and bottom of every flight and be at least as wide and long as the smallest width of the flight. A door opening directly onto a stair would not be acceptable – but the definition of 'stair' in this context is sometimes complicated by the rather complex hybrid arrangements of some conversions (see Appendix 2 item 6, appeal reference 45/3/165).

AD K recognises the hazards presented by doors and doorways relative to stairs, and provisions for minimising risk are shown in Fig. 7.20.

Stair configuration

The provision of stairs has important implications for the layout of both the conversion and the floor immediately below it. In some cases, the need to accommodate stairs will lead to a reduction of the amount of habitable space on the floor below and it is for the householder to decide whether the space gained by the

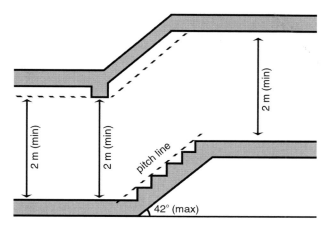

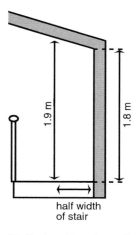

(a) General headroom requirements on stairs

(b) Reduced headroom (loft conversions only)

Fig. 7.19 Stair access: headroom.

conversion justifies the loss of a room, or part of one. However, a number of measures may be considered to limit the amount of space occupied by stairs. Bearing in mind the guidance on minimum pitch and headroom set out in Approved Document K and fire safety requirements in Approved Document B, the following factors may be considered:

Position of stair

To optimise the use of existing circulation areas such as landings, it is practice to run the new stair over the existing staircase where possible (Fig. 7.21). In order to provide appropriate headroom, the size of the ceiling opening required in most cases is only marginally smaller than the overall footprint of the stair.

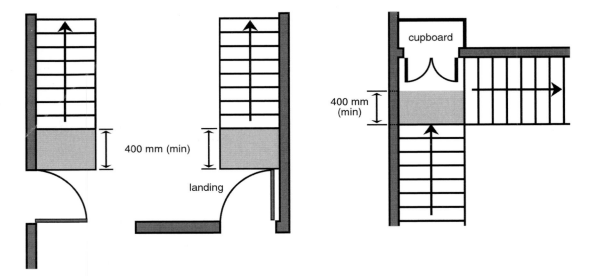

Fig. 7.20   Landings next to doors.

Stair form

A number of configurations are possible. For the purposes of loft conversion, it is assumed that the primary intention of the design process will be to limit the footprint of the stair and the size of the ceiling opening as far as possible. Fig. 7.22 illustrates five basic configurations. As a general principle, complex changes of direction should be made at the bottom of the flight where possible (Fig. 7.22b). Note that while the arrangement illustrated in Fig. 7.22c would usually be acceptable, the provision of winders at the top of the flight introduces increased risk for users.

As already noted, landings must be provided at the top and bottom of every flight, and the width and length of every landing should be at least as great as the smallest width of the flight. The landing may include part of the floor of the building. For example, assuming that the stair illustrated in Fig. 7.22a is 700 mm wide, the clear space at the head and foot of the stair should be 700 × 700 mm (see also Fig. 7.20 with reference to potential obstructions).

Guidance relevant to stairs contained in the Approved Documents (specifically AD B *Fire safety* and AD K section 1 *Stairs and ladders*) is concerned only with safety. Under the guidance, it would be possible to construct a stair that would be entirely impractical for the purposes of moving furniture or large fittings between levels.

Conventional stairs

Approved Document K indicates a maximum pitch of 42 degrees for conventional stairs. By virtue of the guidance, a rise of 220 mm (the maximum permitted) would allow a going not less than 245 mm. There is no regulatory minimum stair

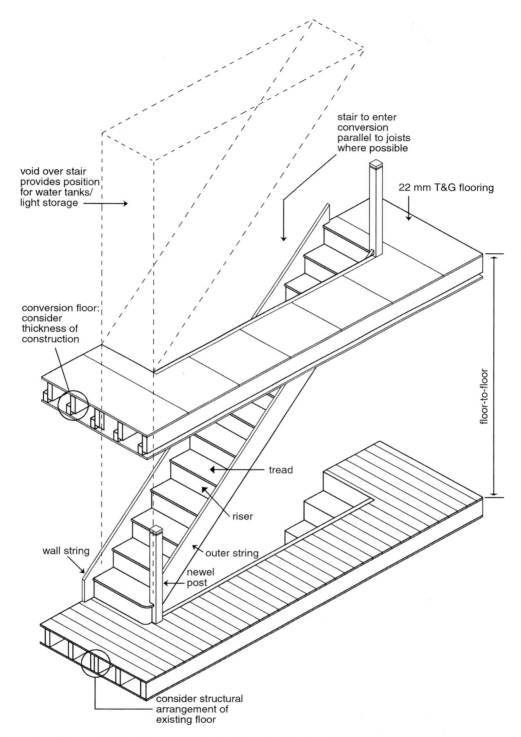

stair to enter
conversion
parallel to joists
where possible

22 mm T&G flooring

void over stair
provides position
for water tanks/
light storage

conversion floor:
consider
thickness of
construction

floor-to-floor

tread

riser

outer string

wall string

newel
post

consider structural
arrangement of
existing floor

Fig. 7.21   Floor-to-floor configuration (balustrade omitted for clarity).

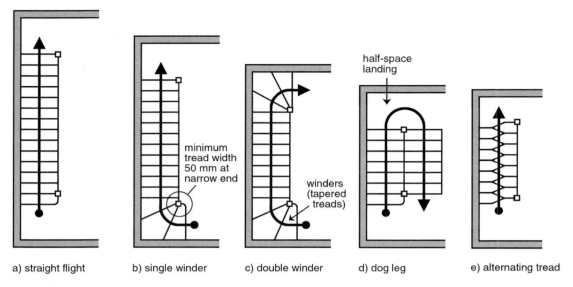

a) straight flight b) single winder c) double winder d) dog leg e) alternating tread

Fig. 7.22 Basic stair forms.

width. 800 mm is preferred; 600 mm would constitute a reasonable minimum. In all cases, appropriate guarding must be provided, and the stair should have a handrail.

Alternating tread stairs

These are sometimes described as space-saver stairs (Fig. 7.23). Maximum rise and minimum going are 220 mm. However, because the going is measured between same-handed treads (i.e. every other step), the maximum rise is, in effect, 440 mm. This gives a pitch slightly in excess of 63 degrees. The advantage with a stair of this sort is its small footprint (Fig. 7.22e). The disadvantage is that safe use is dependent both on familiarity and agility.

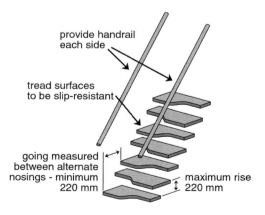

Fig. 7.23 Alternating-tread stair.

Alternating tread stairs must be provided with handrails on both sides and are limited to straight flights or a sequence of independent straight flights. Such a stair may serve only one habitable room together with a bathroom and/or a WC, in which case the lavatory must not be the only one in the dwelling. Note that an alternating tread stair may only be used where it is not possible to accommodate a conventional stair.

Fixed ladders

Guidance on fixed ladders is similar to that concerning alternating tread stairs: a fixed ladder may serve only one habitable room and may only be used where provision of a conventional stair would require alteration to the existing space. Fixed handrails must be provided on both sides of the ladder. Note that a retractable ladder is not acceptable as a means of escape.

Structural implications

Support for stairs is provided in a number of ways, and the structural implications must be considered at an early stage in the design process. In a conventional upper-storey straight flight (open on one side), the outer string is fixed between newel posts on adjacent floors, and the newel posts, in turn, may be fixed to joists or trimmers. The wall string hooks over the landing trimmer, and is either screwed or nailed to the wall.

Additional types of support are required in more complex configurations. Stairs incorporating quarter/half landings, winders (tapered treads) and additional flights, for example, may require the inclusion of load-bearing newels or other means of support at points where the stair changes direction.

As noted elsewhere, it is desirable in all cases for the stair to enter the conversion parallel to floor joists in order to minimise trimming.

Stair provision: practical aspects

Providing stair access is potentially the most problematic aspect of loft conversion. Most difficulties are caused by inaccurate measurements or by a failure to properly assess the relative position of other new elements (such as modified roof slopes) and existing elements (such as the position of doorways that are 'crossed' by the stair). A number of considerations are outlined below:

Floor-to-floor measurement

In most cases, stairs are either purchased off-the-shelf from suppliers or commissioned from specialist manufacturers. Because stairs must have equal risers, it is essential that an accurate floor-to-floor measurement (i.e. the overall rise of the stair) is provided (Fig. 7.21). The safest approach is to fit the conversion's floor joists and trimmers first before making the floor-to-floor measurement: note that the thickness of the conversion boarding (typically 22 mm) must also be included in the measurement. The alternative is to ensure that the conversion floor is constructed to predetermined height relative to the lower floor.

Headroom

As noted above, ensuring adequate headroom is of critical importance. Downstand elements such as beams are potentially problematic. In addition to this, availability of stair headroom beneath a roof slope is sometimes overestimated in the initial survey, particularly if the final thickness of insulation and plasterboard is not properly taken into account. A typical roof slope may require an additional depth of structure slightly in excess of 100 mm (measured at right angles to the rafters) to conform to guidance in Approved Document L. In a roof with a 35-degree pitch, the effect of this would be to reduce headroom, vertically, by approximately 120 mm. Where the headroom infringement is relatively small, it is sometimes possible to create an increased clearance by installing a roof window. In order to reduce the risk of injury, it is of course essential that such a window be hinged at its top rather than the centre.

8 Wall structure

This chapter examines approaches to wall design in loft conversions. Both timber and masonry structures are considered (Fig. 8.1). Construction details, where indicated, represent generally accepted practice, although a variety of approaches may be adopted. In all cases, it is incumbent upon the designer or person carrying out the work to demonstrate, by calculation where necessary, that a proposed approach is suitable for a specific set of circumstances.

EXTERNAL STUD WALLS

New external dormer cheek and face walls in conversions are generally constructed from timber studs and sheathing. This mode of framed construction is widely used when gabled roofs are converted. It may also be used to create a new load-bearing gable end in a building with a hip roof. In this section, the expression 'stud wall' or 'studwork' rather than 'timber frame' is used to differentiate loft wall construction from the conventional version of timber frame used in system-built housing.

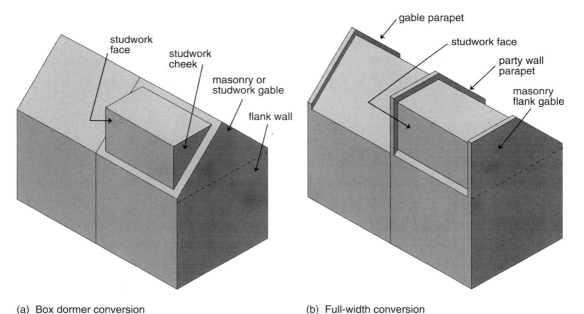

(a) Box dormer conversion (b) Full-width conversion

Fig. 8.1 Basic external wall configuration.

Although similar in some respects to factory produced timber frame construction, the version of timber frame used in loft conversions differs in a number of important ways:

■ External stud walls for conversions are generally built in situ, with timber components cut on site and erected by hand.

■ Tiles, slates or other vertical cladding materials – rather than brickwork – are used to provide a weather resisting finish. External walls in conversions are therefore generally thinner than conventionally constructed equivalents. This is an advantage where it is necessary to create a subordinate structure.

■ Sawn timber is commonly used for loft construction. It is marginally cheaper than surfaced sections (such as ALS/CLS) and its broader sectional area (a nominal thickness of 47 mm rather than the 38 mm of ALS/CLS timber used in conventional timber frame) is more tolerant of nailing inaccuracies. This applies particularly when working blind, for example when fixing external sheathing and battens.

There are a number of advantages in adopting stud wall construction. The materials used are relatively inexpensive and sourcing them is a straightforward matter. Stud frames are constructed swiftly and offer instant load-bearing potential: this means that it is possible to provide a weather-resistant flat roof at a relatively early stage in the construction process. It is also a simple matter to accommodate thermal insulation material between the vertical studs to satisfy current guidance (see also Chapter 10). In addition, the relatively low mass of stud walls is advantageous in older buildings where the strength of existing masonry walls and their foundations is not known.

Stud arrangement and spacing

A grid approach is sometimes adopted in setting out studwork for loft conversions, and this is based on established timber engineering practice with vertical studs at 400 or 600 mm (maximum) centres (Figs 8.2 and 8.3). This has the advantage of minimising the wasteful cutting of sheet materials where these are supplied in 1200 and 2400 mm edge lengths (but see Table 8.1). In addition, it simplifies construction by providing regular and predictable fixing positions. In loft conversions, it is common practice to use 400 mm spacing.

However, the positioning of windows and the need to provide double flanking studs for them means that it is not always feasible to rigorously maintain equal spacing across the entire width of the dormer face and in practice a rather more flexible approach must be adopted. Note also that an axial grid of 400 mm produces a controlling dimension of 353 mm between the faces of 47 mm studs and this inevitably leads to a degree of wastage when insulating material is fixed between them. The following points should also be noted:

■ Factory-produced windows generally do not co-ordinate with a 400 or 600 mm grid. This, and the need to provide double flanking studs, pushes the studwork out of grid and leads to bunching and therefore increased thermal bridging. In

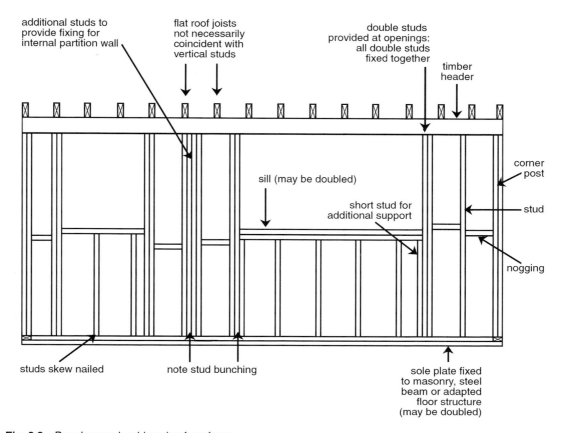

additional studs to
provide fixing for
internal partition wall

flat roof joists
not necessarily
coincident with
vertical studs

double studs
provided at openings:
all double studs
fixed together

timber
header

corner
post

sill (may be doubled)

stud

short stud for
additional support

nogging

studs skew nailed

note stud bunching

sole plate fixed
to masonry, steel
beam or adapted
floor structure
(may be doubled)

Fig. 8.2   Box dormer: load-bearing face frame.

Table 8.1   Dimensions of common components used in external stud wall construction.

Component	Dimensions (mm)
Sheathing boards	**2400 × 1200**, 2440 × 1220, 2500 × 1220, 2697 × 1197, 3050 × 1220, 3050 × 1525
Rigid insulation	**1200 × 2400**
Insulation batts (width)	455, **570**, 600
Plasterboard	**2400 × 1200**
Windows (common nominal widths)	300, 488, 630, 915, 1200, 1342, 1770, 2339
Slate (widths)	200, 250, 300
Plain tile (width)	165

Figures in bold indicate component sizes that co-ordinate with 400 and/or 600 mm stud spacing.
The modularity of cladding (tiles or slates) is generally not considered unless there are a number of closely spaced openings with narrow tile-hanging between them. Note that plain tiles (linear cover 165 mm) are the most commonly used cladding format in loft conversions.

this sense, there is a degree of conflict between the requirement for structural stability and the need to provide insulation. Where bunching is excessive, it may be necessary to provide additional insulation on the inner face of the studwork (see Chapter 10).

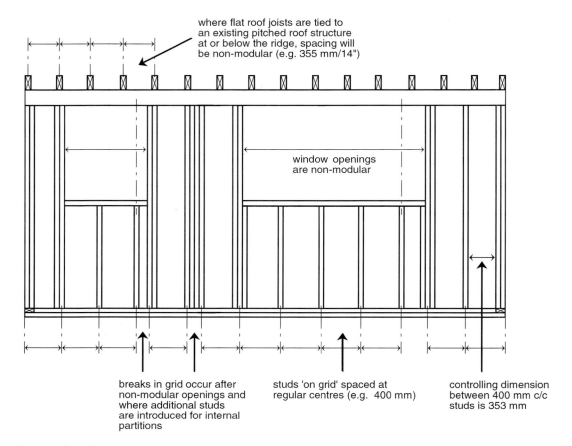

where flat roof joists are tied to
an existing pitched roof structure
at or below the ridge, spacing will
be non-modular (e.g. 355 mm/14")

window openings
are non-modular

breaks in grid occur after
non-modular openings and
where additional studs
are introduced for internal
partitions

studs 'on grid' spaced at
regular centres (e.g. 400 mm)

controlling dimension
between 400 mm c/c
studs is 353 mm

Fig. 8.3   Box dormer: modularity.

- Fenestration was, and is, generally based on the co-ordinating sizes of relatively small masonry units (typically, brick and brick fraction combinations). However, a structural grid based on 400 or 600 mm spacing offers far less subtle arrangements if it is rigidly imposed.
- Closer stud spacing (400 mm rather than 600 mm) is preferable, particularly where heavier vertical cladding materials are to be used such as concrete or clay tiles.
- It is necessary to form a positive nailed connection between the sheathing and the studwork. Corner connections require careful attention. Three corner post configurations are illustrated in Fig. 8.4.

Elements of stud wall construction

Timber studwork loft walls are generally constructed from C16 sawn softwood. Timber elements are cut square, butt jointed and nailed together (Fig. 8.6). Framing anchors may also be used in this application. The combination of reliable sheathing materials and ballistic nailing equipment means that housed joints are no longer widely used.

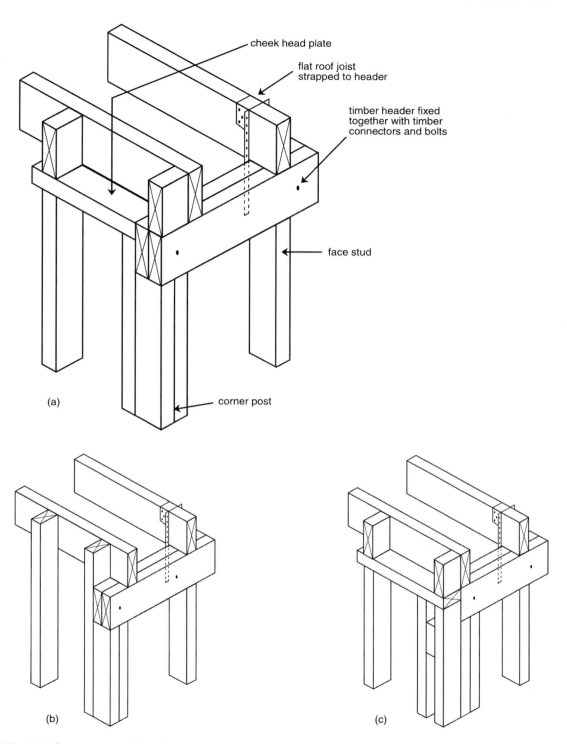

cheek head plate

flat roof joist
strapped to header

timber header fixed
together with timber
connectors and bolts

face stud

(a)

corner post

(b)

(c)

Fig. 8.4   Corner post configurations.

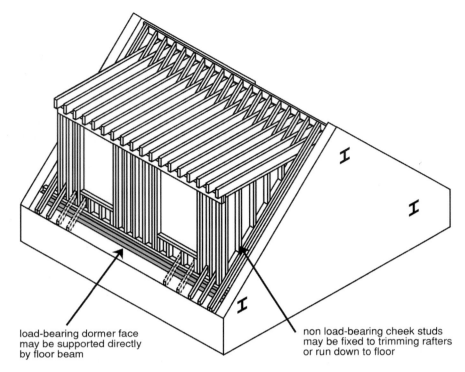

load-bearing dormer face
may be supported directly
by floor beam

non load-bearing cheek studs
may be fixed to trimming rafters
or run down to floor

Fig. 8.5 Box dormer studwork.

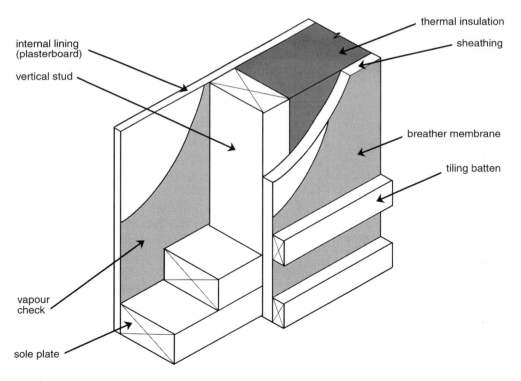

thermal insulation

sheathing

internal lining
(plasterboard)

vertical stud

breather membrane

tiling batten

vapour
check

sole plate

Fig. 8.6 External stud wall: basic construction.

In the examples shown in Figs 8.2 and 8.5, dormer face studs are skew (slant) nailed at the base to sole plates and fixed at their upper extremities to the header. Note that in most box dormer configurations, the face studs support the header and must therefore be capable of supporting a flat roof load.

Studwork cheek walls are generally not configured for roof loading (Figs 8.1a and 8.5). However, where built directly off a structural floor, they may be used to support the edges of the trimmed opening in the original roof slope. Cheek studs are fixed to a single head plate at the top (Fig. 8.4a) or run to full roof height and fixed to the flat roof joists (Fig. 8.4b).

Terminology

There is a degree of regional and local variation in the terms used to describe stud wall components. There are few cut and dried definitions. Where a number of different terms are used to describe a given component, the most commonly accepted expression is used here. Variant names are also provided.

Sole plates

Sole plates (sometimes also called plates, sill plates, floor plates, base plates, sills/cills or bottom rails) provide a fixing for the feet of vertical studs (Fig. 8.2). The sole plates themselves may be fixed to existing masonry, beams or to a new structural floor. The effectiveness of the sole plate-to-substrate fixing must be considered in relation to structural wind loads, including the effects of wind uplift on the relatively low-mass structure. Both single and double sole plates may be used depending on loading conditions and the structural elements to which they are fixed.

Sole plates may be fixed in a number of ways depending on the design of the conversion. These include nailing or heavy duty screwing to joists in the new floor structure, bolting to steel floor beams, shot-firing to steel beam flanges or fixing directly to existing masonry with straps, heavy duty expansion fittings or chemical anchors (Fig. 6.14). The use of expansion fittings requires particular care with older and often friable brick masonry, and strapping or chemical anchoring should be considered. Shot-fired fixing of timber to steel is well established in US timber frame construction and is becoming standard practice in the UK.

Studs

Sawn softwood studs with minimum dimensions of 100×47 mm are set at 400 or 600 mm (maximum) centres to support anticipated loads and to conform to the dimensions of sheet materials. Where they are to be load-bearing, wall studs are designed as compressive members with a minimum strength class of C16. Studs are skew nailed with at least two nails for each joint.

The choice of external cladding may also influence stud spacing, particularly where heavier materials are to be used. Note that it is the studs, and not the sheathing alone, to which tiling battens must be fixed.

Under no circumstances should structural studs be cut into in order to provide a housing for noggings, sills, lintels or other horizontal members. Where additional support for these is required, short studs or cripple studs should be provided.

Noggings

Fixing one row of staggered noggings at half stud height is recommended in load-bearing walls (Fig. 8.2). Noggings that have the same sectional dimensions as studs are introduced where there is an increased need for rigidity and to resist buckling. They are also used to provide support for heavy internal fittings such as radiators and sanitary appliances. Noggings may also be provided at the sheet edges of wallboards and sheathing to provide fixing positions. They are sometimes described as noggins, nogs or dwangs.

Header

The header is the uppermost horizontal element of a load-bearing dormer face frame in a box dormer conversion and it generally provides a bearing for flat roof joists (Figs 8.2 and 8.4). This component is variously described as a head beam, header beam, window header, top head, header joist, head plate or double head-plate beam.

The header is generally formed from a pair of deep-section joists but, as with other elements of horizontal load-bearing structure, it must be designed by a structural engineer: most local authorities require that calculations be submitted. In box dormer conversions, however, it should be noted that the header seldom spans clear at full length, and face studs generally provide intermediate support. Contrast this with the head beam (generally steel or flitch) provided at the face in a masonry flank gable conversion which in some cases can be configured as a single clear span (Fig. 8.9).

The advantage of using a header which has a considerably deeper section than the rail/binder combination used in timber frame construction is that it provides a ready-made and continuous lintel with an acceptable head height for window openings. It is also robust enough to support a new flat roof structure and at the same time allows a degree of offsetting of point loads. This frees both designer and builder from the need to meticulously line up each flat roof joist with a vertical stud. Equally, it offers a degree of freedom in positioning the vertical studs, provided that they are not more than 400 mm apart. The ability to make such changes contingent upon conditions that arise during a conversion considerably speeds up the construction process.

External sheathing

Sheathing boards (sometimes called bracing boards), usually exterior sheathing-grade plywood or oriented strand board (grades OSB3 or OSB4), are nailed to the external faces of the studs. In conventional timber frame construction, sheathing

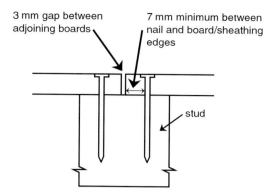

Fig. 8.7 External sheathing: fixing to studwork.

boards may be as little as 8 mm in thickness; in loft conversions, 12 mm, 15 mm and sometimes 18 mm boards are used.

The primary function of sheathing is to stiffen the wall and provide resistance to racking deformation. Sheathing also assists in reducing wind penetration and provides a supporting background for both breather membrane and insulation materials.

Nail specifications and nailing schedules are influenced by the thickness of the sheathing material and loading considerations. However, with plywood or OSB up to 12 mm thick, the use of 3 mm diameter corrosion-resistant 50 mm nails would be reasonable. Boards are nailed to studs and horizontal members at 150 mm centres along their edges and at 300 mm intervals mid-panel. A gap of 3 mm is allowed between adjacent boards (Fig. 8.7).

Board sizes vary depending on the country of origin. A proportion of plywood used in the UK is manufactured in North America and in many cases delivered boards measure 2440 × 1220 mm (i.e. 8' × 4') and will therefore require cutting to conform to a metric structural grid. Equally, stud positions can be adjusted to accommodate the additional length, but note that internal wallboards conform to metric dimensions (1200 × 2400 mm).

Breather membrane

It is practice to cloak the external face of the sheathing with breather membrane. There are a number of proprietary versions but all of them work in a similar way, allowing water vapour to escape from the structure whilst preventing any wind-driven rain that has penetrated the tile/slate cladding from entering it. In addition, breather membrane is an effective barrier to wind penetration.

Breather membrane is fixed to sheathing with stainless steel staples. A 100 mm lap should be made at horizontal joints with the upper sheet always overlapping the lower. At vertical joints, the lap should be a minimum of 150 mm.

The use of roofing underlay or felt rather than breather membrane is generally considered to be unsatisfactory. Traditional roofing underlay has low vapour permeability and therefore increases the risk of interstitial condensation either on the sheathing or within the wall structure. Guidance provided in Approved Document C *Site preparation and resistance to contaminants and moisture* indicates that cladding should be separated from insulation or sheathing by a vented and drained cavity, with a membrane that is 'vapour open'.

Openings

Where openings for windows are to be formed, additional flanking studs should be provided at each side of the opening (Fig. 8.2). It is practice to provide a sill or double sill beneath window openings. Lintels, where used, are supported by cripple studs. In most small to medium-sized conversions, however, the header acts as a lintel.

Supporting structural steel in stud walls

Appropriately designed, timber frame walls are capable of providing support for structural steel components such as roof beams. The studs, sheathing and supporting structure below must be correctly adapted if the wall is to fulfil this function. Generally, a timber post is built into the stud wall and this may be either a solid timber section or a number of studs fixed together. Structural engineering calculations are required to prove the adequacy of the beam and the timber post and structure that will support it.

VERTICAL CLADDING

Treated timber battens for vertical tile or slate hanging are fixed on top of breather membrane. The gauge (vertical spacing between horizontal battens) is primarily governed by the choice of cladding material (Table 8.2). The positions of window openings and other fixed points such as the junction at the bottom of the wall and the soffit/fascia at the top must also be taken into account.

In all cases, battens should be at least 1200 mm in length and fixed at a minimum of three supporting points. Butt joints should be staggered and splay-cut joints cut at 45 degrees. Care should be taken to ensure that battens are nailed through the sheathing and directly into the studs – nailing to sheathing alone would generally not be an acceptable means of support. Where battens are cut, they should be sawn cleanly and the ends well treated with timber preservative.

Table 8.2   Vertical cladding.

Cladding material	Unit dimensions (mm)	Batten size (mm)	Typical gauge (mm)
Plain tiles	265 × 165	38 × 25	114
Slate (ladies)	400 × 200	50 × 25	155*
Slate (countesses)	500 × 250	50 × 25	205*

* 65 mm head lap assumed.

In exposed locations where there is an increased likelihood of wind-driven rain penetrating joints in the cladding, vertical counter battens should be fixed behind the horizontal ones to create a void from which any water may drain away freely and within which air may circulate. Section 5.27 of Approved Document C *Site preparation and resistance to contaminants and moisture* (2004) indicates that where cladding is supported by timber components, the space between the cladding and the building should be ventilated, although the use of counter battens is not specifically referenced.

Note also that section 5.32 of the same Approved Document specifically recommends the use of 25 mm checked rebates for window and door reveals in areas of high exposure to driving rain.

FIRE RESISTANCE OF DORMER STUD WALLS

The fire resistance of an external wall becomes of critical importance relative to its relationship with a relevant boundary. Whilst a normally constructed solid masonry wall (225 mm) or a cavity masonry wall (brick 100 mm, cavity 50 mm, block 100 mm) is generally assumed to exceed the required fire resistance for the purposes of a loft conversion, the same assumption cannot be made about an external timber stud wall.

Fire resistance and space separation are examined in detail in Approved Document B *Fire* (sections 13 and 14). In simple terms, however, where a stud cheek wall is within 1000 mm of a relevant boundary it is generally considered to constitute an 'unprotected' area and this is generally the case irrespective of the vertical cladding material used. Where such a wall has an area greater than $1\,m^2$, it requires 30-minute fire resistance to both sides (Fig. 8.8).

The following measures are generally taken to satisfy the guidance set out in Approved Document B:

■ External lining: a fire resistant board providing 30-minute fire resistance is fixed on top of the sheathing before breather membrane, battens and tiles are fixed. Proprietary fire resistant boards are produced by a number of manufacturers and are generally made from calcium silicate.
■ Internal lining: a generally accepted solution is to fix a 12.5 mm plasterboard lining with scrimmed joints and a gypsum plaster finish.

MASONRY WALLS (EXTERNAL)

Brick masonry may be used in the construction of gable, flank and party walls in loft conversions. However, two points are emphasised:

(1) The practice of building up gable or flank walls is not universally accepted by local authority planning departments under permitted development (see Appendix 3).

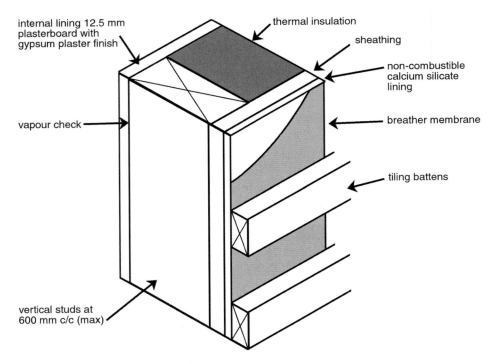

internal lining 12.5 mm plasterboard with gypsum plaster finish

thermal insulation

sheathing

non-combustible calcium silicate lining

vapour check

breather membrane

tiling battens

vertical studs at 600 mm c/c (max)

Fig. 8.8 External stud wall: fire resistance.

(2) It is necessary to invoke the Party Wall etc. Act 1996 where work on a party wall is proposed (Chapter 1).

The use of brickwork for gables and flank walls is often advantageous for a number of reasons. Building up a wall in brick similar to the original may be aesthetically preferable to tiled, slated or rendered stud walls. This is particularly the case in end-of-terrace and semi-detached conversions where exposed flanks and gables are a dominant feature. Indeed, in some conservation areas, the use of masonry may be a pre-condition for obtaining planning approval.

In terraces where a number of conversions are to take place, the use of masonry walls between conversions introduces a degree of vertical discipline and goes some way to eliminating the often ragged and unsightly junctions that occur where stud-built conversions abut.

Brick masonry flank walls, unlike stud walls in most cases, transmit their self-load directly to the foundations. This may simplify floor design to some extent. In addition, masonry flanks provide a potential load path for some flat roof loads that might otherwise require support from the floor or floor beams (Fig. 8.9). The following points should also be noted:

- Brick masonry, both cavity and solid (225 mm), offers good resistance to fire and generally requires no additional detailing to achieve this.
- Correctly detailed, a masonry wall requires little or no maintenance during the life of the building.

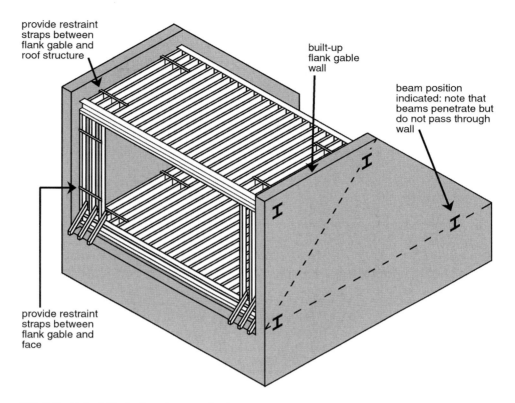

provide restraint
straps between
flank gable and
roof structure

built-up
flank gable
wall

beam position
indicated: note that
beams penetrate but
do not pass through
wall

provide restraint
straps between
flank gable and
face

Fig. 8.9 Full-width dormer: masonry flanks.

- Constructing a masonry gable or flank gable wall, rather than building a floor-supported inset stud cheek, maximises the amount of useable space inside the conversion.
- The relatively high mass of brick masonry makes it an effective barrier to the transmission of sound. Loft party walls constructed in solid 225 mm brick masonry are likely to provide adequate sound insulation between dwellings without significant modification.

Against these advantages must be weighed the relatively higher material costs of building in masonry and the fact that brick construction is generally more time-consuming and therefore costly than erecting studwork (although this does not apply to walls built in solid blockwork – see below). Equally, providing adequate thermal insulation would need consideration where solid brick masonry walls or traditional narrow (50 mm) cavity walls are to be built (see Chapter 10).

Consideration should also be given to the physical condition of existing party / external walls, and the possible structural consequences of increasing the masonry loading on them, before undertaking work. This point is of particular importance where the new brickwork loading will be unevenly distributed, for example, where a gable wall is extended outwards towards the rear of the building to form a new flank gable for the loft conversion.

Hip-to-gable conversion

For houses with hipped roofs, building up a new gable wall to replace the original roof hip makes it possible to support a new flat roof and floor beams. This process is generally described as a hip-to-gable conversion. Note that such conversions may be carried out on both solid and cavity masonry walls. It is of course also possible to form a gable in studwork.

A variant approach, which may equally be applied to dwellings that already have gables, is to build up the rear portion of the flank wall to the height of the new flat roof deck. Where the dwelling to be converted has a party wall parapet, that is, a parapet that is higher than the roof covering and ridge, the extended flank wall will also be higher than the level of the flat roof (Fig. 8.1b). It must therefore be detailed with an appropriate coping and flashings to the new flat roof (see also *Parapet walls* below).

Safety considerations during construction

It is stressed that the process of constructing gable and flank walls is uniquely hazardous in loft conversions. By their very nature, gable walls do not enjoy the same degree of resistance to lateral loading as ordinary walls, which are buttressed by returns at each end. The same also applies to flank gables in many cases. During construction, such free-standing walls are highly vulnerable to wind loading and accidental impact, with the consequent risk of collapse.

In new build, it is practice to erect roof trusses *before* gables are constructed. The gable is progressively strapped to the roof structure as the masonry is brought up, and a margin of safety is thus maintained. This approach, of course, is rarely possible in a loft conversion. For this reason, the use of shoring and strutting must be considered, to ensure the stability of the new brickwork during the course of construction. This should remain in place until the new wall is strapped to bracing roof and wall elements.

Lateral restraint of flank gable walls

Any design that incorporates new masonry flank gable walls must take account of the need to provide adequate lateral restraint for them. While Approved Document A *Structure* does not provide guidance specific to loft conversions with masonry walls and flat roofs, section 2C provides general guidance on buttressing for walls and lateral support for gables.

In the case of flank gable walls, a feature of most designs is that the wall is generally not returned at its end and thus it is not buttressed in the normal way. Other positive methods of restraint for such a wall must therefore be provided (Fig. 8.9). The form of such restraints is dependent on the design of the conversion, and any system of restraints must be designed and specified by a structural engineer or other competent person. The following elements may be considered:

- *Floor level*: restraint provided by tension straps from wall to floor structure (see Chapter 7)

- *Wall*: provision of tension straps tying masonry flank to face studwork
- *Roof level*: provision of tension straps tying head of flank to flat roof joists

In most cases, flank gable walls are raised in solid masonry. Approved Document A does not provide guidance on fixing tension straps to solid masonry walls, but Fig. 7.16b illustrates one possible method of achieving this.

Brick selection and size

Before undertaking any conversion where new brickwork is to be used in conjunction with existing walls, careful measurements of the existing bricks and brickwork should be made and a reliable source of matching bricks, either new or reclaimed, identified.

It should be noted that bricks with standard dimensions did not appear until 1904 when the Royal Institute of British Architects (RIBA) adopted the Southern Brick Standard with a brick height of $2\,^5/_8$" (66.8 mm). Until 1904, there were only popularly-used sizes of bricks and even these were subject to considerable local and regional variations.

Other standards were introduced during with twentieth century. RIBA introduced an additional standard – the 'northern' brick standard – of $2\,^7/_8$" (73 mm) in the 1920s. As their names suggest, the 'northern' and 'southern' standards reflected regional usage. However, these bricks were standard only in name, and local variations continued until the introduction of the imperial standard brick in 1965 with measurements of $8\,^5/_8$" $\times\,4\,^1/_8$" $\times\,2\,^5/_8$" (219 × 104.8 × 66.8 mm).

The current British Standard metric brick was introduced in 1974 and is 215 × 102.5 × 65 mm. However, some imperial-dimensioned bricks are still manufactured (see below).

Reclaimed bricks

There is a degree of cachet associated with the use of reclaimed bricks, and these are now used extensively in hip-to-gable conversions, particularly in pre-1914 buildings. Local or stock bricks are routinely recovered during demolition work and in many areas these provide a source of well-matched bricks in terms of colour, composition and size. Bricks are cleaned and palletised for re-use and are generally supplied in batches of 500. Their relatively high cost, generally three to four times that of common bricks, reflects the labour-intensive nature of the recovery process.

It should be noted that the physical characteristics of individual bricks in reclaimed batches may vary considerably and it is unlikely that a supplier will be able to provide details of, for example, the compressive strength and frost resistance of the bricks. The latter is of importance where the wall is to be carried above the roof decking as a parapet. The potentially limited crushing strength of such bricks is an additional consideration, particularly where the wall to be constructed is intended to support a beam. Careful attention should be given to the specification of padstones or bearing plates in such cases (see also Chapter 6).

Laying reclaimed stock bricks in a satisfactory manner is an important additional consideration. Dimensional irregularities are a feature of such bricks, and

Fig. 8.10   Hip-to-gable conversion in stock brick.

achieving a visually acceptable result is a test of the bricklayer's craft (Fig. 8.10). The use of cement:lime:sand mortar, rather than cement:sand, is recommended for reclaimed bricks. With softer stocks, consideration should also be given to the use of a designation iv (weaker) mortar to reduce the risk of the brickwork cracking.

New bricks

There are approximately 1200 different types of bricks available in the UK and it is often possible to match, both in appearance and size, new bricks with old. As well as their lower cost, there are a number of advantages in specifying new rather than reclaimed bricks. Their characteristics are declared (for example, crushing strength, frost resistance and the presence of soluble salts) and, in addition, there is generally little wastage due to selection.

Many brick manufacturers make and hold bricks in various imperial measurement sizes. The most common imperial-compatible new bricks are (in height): 80 mm, 73 mm ($2\,^7/_8$"), 67 mm ($2\,^5/_8$") and 50 mm (2"). The majority of these bricks are 215 mm long and 102.5 mm wide.

It is rarely possible to achieve a perfect tonal and textural match with any new brickwork and a degree of visual discontinuity is inevitable, even if it is only the jointing that stands out. It is possible to harmonise the appearance of mismatched brickwork by tinting. British Standards exist for the pigments used in tinting, but there are currently no standards specifying tinting procedures.

Where suitable new or reclaimed stocks are not available, bricks to match may be made to order. A number of brick manufacturers are able to do this, although for small batches, less than 1200 for a typical solid masonry hip-to-gable conversion, this may not be considered economical.

Solid blockwork

Gables or flank gables may also be built up with lightweight aerated concrete blocks that are the same width as the wall (Fig. 8.11). This approach may be adopted in buildings with either existing solid or cavity walls. Lightweight blockwork has the advantage of offering a relatively high level of thermal insulation, a relatively low mass and, because individual units are comparatively large (standard face dimensions of blocks are 440×215 mm), walls may be constructed relatively swiftly. A day lift should not exceed six courses (say 1350 mm).

Note that a blockwork wall requires a final weather-resisting finish of render or cladding and will usually require supplementary internal insulation to meet current guidance (see Chapter 10).

Aerated concrete blocks are produced in a range of sizes that co-ordinate reasonably well with most existing walls, cavity or solid, and are generally available in 200, 215, 255, 265, 275 and 300 mm thicknesses. Because they have a relatively low compressive strength, they should be laid with one of the weaker mortars: designation iii is generally used in this application. In the interests of achieving a better bond, some manufacturers recommend that blocks be laid in cement:lime:sand mortar. Blocks should be laid with a regular bond pattern with a minimum overlap of a quarter of a block. Mortar joints, both inside and out, should be left recessed to provide a key for render and plaster finishes.

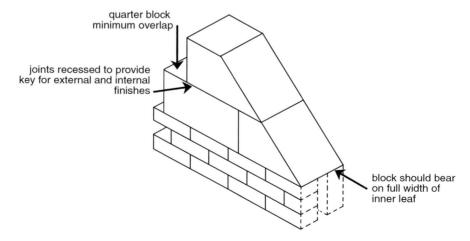

quarter block minimum overlap

joints recessed to provide key for external and internal finishes

block should bear on full width of inner leaf

Fig. 8.11 Full-width aerated concrete blockwork for new gable wall.

Note that whilst standard $440 \times 215 \times 100$ mm blocks can be laid face down to produce a 215 mm wall, this approach would result in an undesirable increase in thermal bridging. It is also stressed that new blockwork should be bedded across the full width of the existing inner (load-bearing) leaf in a building with a cavity wall.

Mortar and brickwork

In low-rise buildings, mortars of designations iii and iv are generally used, depending on brick type and exposure. Note, however, that Approved Document A contains only guidance related to mortars of designation iii or stronger.

In cement:lime:sand mixes, ordinary Portland cement and dry hydrated lime (calcium hydroxide) are generally used. Hydrated lime improves workability, provides a better bond and offers enhanced water resistance. Sand is now generally described as fine aggregate in guidance. Where cement:sand mortar is used, it is generally mixed with an air-entraining agent (plasticiser) to improve workability.

The higher the proportion of cement (binder) in a mortar mix, the higher its compressive strength will be. However, because strong mortars are less flexible, they are more likely to cause destructive localised stresses. For this reason, mortar should generally have a lower compressive strength than the bricks it is designed to bond.

This is of particular importance when using reclaimed bricks of unknown crushing strength. It should also be noted that the strength of brickwork is determined by both bricks and the choice of mortar: it is generally the lower strength of mortar, rather than the higher strength of bricks, that determines bearing capacity.

Where frogged bricks are to be used, it is good practice to lay them frog up. This reduces localised stresses within the brickwork, particularly on brick edges where the wall is designed to support structural elements such as beams. The wall will also have better sound insulation properties because there are less likely to be voids in the brickwork.

It should also be noted that using imperial brick sizes in conjunction with metric blockwork, as is occasionally the case with hip-to-gable conversions in inter-war houses, will lead to misaligned bed joints caused by the relatively greater thickness of pre-metric bricks. When tying the inner and outer leaves, therefore, it is practice to use a two-part wall tie with vertical channel section and sliding tie to accommodate differences in bedding height (Fig. 8.12). An alternative is to use helical screws. Under no circumstances should conventional wall ties be bent to accommodate differences in bedding height. Note that Approved Document A now specifies the use of stainless steel ties in all domestic dwellings.

Parapet walls in loft conversions

Rather than terminating masonry flank walls beneath the roof covering, the walls may project above the roof surface to form a parapet. Guidance on maximum

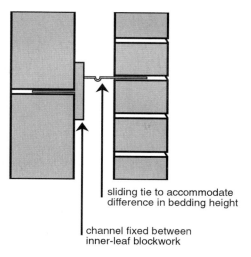

sliding tie to accommodate
difference in bedding height

channel fixed between
inner-leaf blockwork

Fig. 8.12   Two-part wall tie.

heights and minimum thicknesses of parapet walls is provided in Approved Document A *Structure*. In loft conversions, though, achieving the *minimum* height of a parapet is often a significant consideration (Figs 8.13a and 8.13b).

Where there is a need to form a particularly low parapet wall relative to the finished roof deck, the requirement to form a weather-resisting junction between the roof covering and the parapet wall becomes the limiting factor. Note that the parapet height must be referenced to the highest point of the flat roof, taking firring and the thickness of roof coverings into account. Over a typical loft conversion flat roof span of 3 m, the fall of the flat roof will generate a height difference of 75 mm, or about one brick height, between the highest and lowest points.

In most cases, the minimum parapet height that can be achieved for practical purposes relative to the highest point on a flat roof is about 225 mm (Fig. 8.17) plus the thickness of the coping or capping which could be coping stones, brick-on-edge or sheet metal dressed over boards. In all cases, the minimum upstand for the roof covering should be considered. This is generally 150 mm and is intended to provide a degree of protection against rainwater splash. It also limits the risk of damage caused by temporary inundation.

It should be noted that parapet walls are exposed to a relatively high degree of weathering, and particular care should be taken in their detailing to ensure durability. Typically, both sides of parapet wall brickwork are exposed. However, in a low level parapet (described above), exposure is reduced somewhat on the deck side of the wall, and brick choice is less of a critical factor than it would be in a relatively high parapet. In all cases, however, using a mortar of designation i or ii would be reasonable.

In order to protect the brickwork, the coping or capping (whether masonry or metal) should oversail the parapet wall on both sides. Note that the damp proof course (dpc) at cover flashing level is placed *above* the cover flashing.

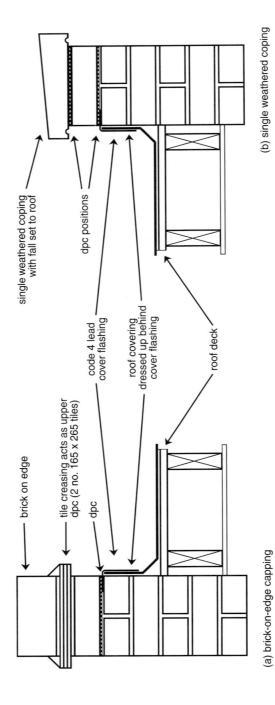

(a) brick-on-edge capping

(b) single weathered coping

Fig. 8.13   Low-level parapet walls in solid masonry.

INTEGRATING NEW ELEMENTS WITH OLD

Junctions between existing roof structures and new walls are inevitable given that most loft conversions are subordinate structures. Angular relationships between new elements of structure and the existing roof are frequently complex, and providing an appropriate weatherproof interface is seldom a straightforward matter (Fig. 8.14).

Weather-resisting junctions are generally formed from lead sheet. Lead is malleable and this means that as well as being relatively easy to work, at least in simple applications, it is capable of accommodating small movements between elements of structure over time without fracturing. In addition its considerable weight even in lighter codes (20.41 kg/m^2 for code 4 sheet) means that it is also resistant to wind uplift when appropriately clipped.

Junctions between single planes (i.e. wall-wall, wall-roof slope) are the easiest to form. Fig. 8.15 illustrates a side abutment with lead step flashing between a chimney stack and a new dormer cheek before tilehanging. Note that the brick joints are raked clean with all mortar residue removed. Each step of the flashing is folded over by 25 mm for fixing into the joint. The lead flashing is fixed in position with a lead wedge for each step before the joints are filled with mortar or proprietary lead sealant.

The junction between an existing roof slope and a new tiled dormer cheek is weatherproofed using soakers, usually in a lighter code (code 3). Soakers are interleaved between the tiles of the existing roof slope and turned up behind the battens (but on top of the underlay) of the studwork cheek (see also Fig. 9.14).

The junction between a tile-hung dormer face and an existing portion of roof slope is generally provided by a cover flashing that is dressed over the lower tiles and fixed behind the battens (and underlay) of the vertical dormer face. The cover flashing should be in at least code 4 lead, and it should be clipped to resist uplift. The use of handed 90-degree corner tiles provides a weatherproof junction between the dormer face and cheek (Fig. 8.16).

Fig. 8.16 also illustrates the sort of multi-plane junction that is common in loft conversions. In this case, the dormer face and dormer cheek intersect with the original roof plane. In order to adequately weatherproof a junction of this sort, lead welding (lead burning) or bossing is generally necessary to allow the creation of appropriately shaped flashings.

Similar junctions may occur where parapet walls form a junction with other roof elements. Fig. 8.17 illustrates a junction between two parapet walls, one with coping, one capped with lead sheet, and the kerb-upstand of an asphalt roof. Whilst weatherproofing junctions of this sort is a relatively time-consuming process, it should be noted that most roof failures occur at junctions between planes and not within the planes themselves.

Fig. 8.14 Flashings and weatherings: face to slope.

Fig. 8.15 Cheek to stack.

Fig. 8.17 Parapet treatments.

Fig. 8.16 Cheek to slope.

Chimney cowls

The capacity of chimneys to allow the introduction of moisture directly to the fabric of a dwelling is sometimes overlooked. This is, in part, because flue-offsets in the roof space intercept the bulk of the precipitation, which diffuses into the brickwork unnoticed until the loft is occupied. Given that a typical four-pot stack is capable of directly admitting in excess of 30 gallons of water each year, consideration should be given to the provision of appropriate cowlings.

COMPARTMENT (PARTY) WALLS

The wall separating adjacent but separately occupied properties in a terrace or semi-detached dwelling is generally referred to as a party wall (the term has a distinct legal meaning as well – see Chapter 1). For the purposes of fire safety, such a wall is described as a compartment wall.

Until the late nineteenth century, adjacent terraced dwellings were not routinely separated from each other by compartment walls in the roof void: walls were sometimes built up only to the level of the upper-storey ceilings. Separation may be achieved by raising a new compartment wall, at full thickness, in the roof space. A 215 mm masonry wall with bricks laid frog-up will generally conform to guidance on sound and fire resistance. The following points should also be noted:

- Where a new party wall is to be constructed, it will be necessary to invoke the Party Wall etc. Act 1996 and to inform the adjoining owner or owners. For more details, please refer to Chapter 1.
- A compartment wall should be taken up to meet the underside of the roof covering or deck, with fire-stopping where necessary at the wall/roof junction to maintain the continuity of the fire resistance. The compartment wall should also be continued across any eaves cavity.

Timber components such as beams, joists, purlins and rafters may be built into or carried through a masonry compartment wall provided that the openings for them are kept as small as possible and fire-stopped. Similarly, roofing felt and tiling battens, if fully bedded in mortar, may be carried over a compartment wall. Guidance to this effect is provided in Approved Document B. However, the practice of 'building in' timber structural elements is generally discouraged except where no alternative exists, as in the case of tiling battens that must bridge the compartment wall in a terrace.

Mortgage lenders generally require that separation is present in the roof void and, for this reason, it may be found that separation has been provided subsequent to construction by building up the party wall in 100 mm blockwork or similar. This is unlikely to provide adequate sound insulation or a suitable bearing for new floor and roof beams, and replacement with solid brick masonry may be the only suitable solution.

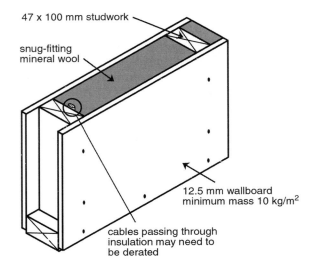

47 x 100 mm studwork

snug-fitting
mineral wool

12.5 mm wallboard
minimum mass 10 kg/m²

cables passing through
insulation may need to
be derated

Fig. 8.18   Partition wall sound resistance.

INTERNAL PARTITIONS

Partition walls within the conversion provide separation between rooms and may be framed in 100 × 47 mm timber studwork and clad with plasterboard. Partitions must be fire resisting in certain circumstances (see Chapter 4).

In addition, internal walls between a bedroom or a room containing a water closet, and other rooms, must provide reasonable resistance to airborne sound. This may be satisfied by adopting the detail illustrated in Fig. 8.18. Sound resistance measures need not apply if an internal wall contains a door, or if an internal wall separates an en suite WC and the associated bedroom.

WINDOW AND DOOR SAFETY

Windows present two types of risk: the danger from falling and the risk of cutting and piercing injuries caused by broken glass. Guidance on minimising these related but distinct hazards is set out in Approved Document K *Protection from falling, collision and impact* and Approved Document N *Glazing – safety in relation to impact, opening and cleaning*.

Windows

Glazed areas that are less than 800 mm above the finished floor are considered to be in a critical location for the purposes of Approved Document N. The most widely-used method of conforming to the guidance is to provide glazing that meets safe breakage criteria set out in BS 6206: 1981.

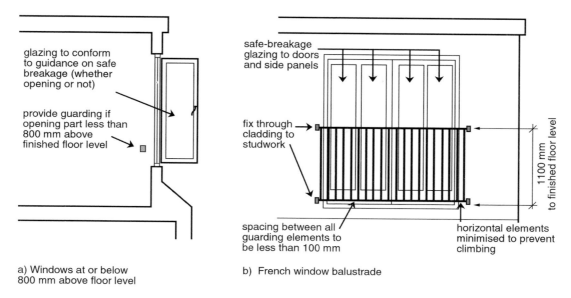

glazing to conform to guidance on safe breakage (whether opening or not)

provide guarding if opening part less than 800 mm above finished floor level

safe-breakage glazing to doors and side panels

fix through cladding to studwork

1100 mm to finished floor level

spacing between all guarding elements to be less than 100 mm

horizontal elements minimised to prevent climbing

a) Windows at or below 800 mm above floor level

b) French window balustrade

Fig. 8.19   External openings: guarding and glazing.

To minimise the risk of falling, the guidance in Approved Document K is that guarding be provided to opening windows at or less than 800 mm above floor level (Fig. 8.19a). However, guarding is not required for a means of escape window set in a roof slope even if it is less than 800 mm above floor level.

Balustrades

Inward opening French windows are a popular feature in loft conversions. These must be protected by a balustrade of appropriate dimensions that is fixed either to structural studwork or to masonry (Fig. 8.19b). The balustrade protecting a French window must be a minimum of 1100 mm above finished floor level and should have only a minimum of horizontal elements to prevent it being climbed.

In addition, the spacing between all vertical and horizontal elements in the construction of the balustrade, including ground clearance, should be such that it would prevent a sphere of 100 mm diameter passing through the structure. Additional guidance on guarding heights is provided in Approved Document K.

Glazing requirements for doors

Doors present additional hazards. The guidance for glazing in doors (such as French windows) is therefore more rigorous than that for windows: glazed door elements 1500 mm or less above floor level must conform to guidance on safe breakage. In addition, glazed side panels within 300 mm should be provided with safety glazing.

9 Roof structure

This chapter considers commonly encountered roof forms and the modifications that may be made to them as part of a loft conversion. Dates are provided for guidance only and are intended to indicate the likely use of a roofing system for a house constructed during a given period. However, roofs of any era may be hybrid structures that do not necessarily conform to textbook examples. A full understanding of the structural function of the roof must be gained before any alteration is undertaken.

ROOF TYPES

The cut roof (common to about 1950)

The term 'cut roof' is a fairly loose one and is used to distinguish the traditionally constructed timber roof from the trussed rafter form that has now largely superseded it. In the cut roof, timber components such as rafters, ridge, struts, purlins and ceiling joists, were sawn on site and nailed together to create a roof structure (Fig. 9.1a). This system of construction was almost universal in the UK until World War II. Generally, the structural design of such roofs was based on custom rather than calculation: it is fortunate that in many cases (but not all), cut roofs are reasonably tolerant of modification.

The sometimes generous dimensions of traditional cut roof components and the relatively small number of projections (e.g. struts and collars) within the roof void mean that these roofs are generally the easiest to convert, although a new structural floor is required in most circumstances.

Cut roofs often depend upon a load-bearing internal wall for some of their structural integrity. This, in itself, is often advantageous as far as loft conversions are concerned.

The TDA roof truss (common 1947–1980)

Loft conversions in TDA timber truss roofs are generally feasible provided there is adequate headroom, but they are generally more troublesome than those in the cut roof described above. The principal difficulty lies in positioning new structural members between the struts, ties and hangers present in the roof void. However, the TDA trussed roof is rather more robust in its construction and provides more room for manoeuvre than the trussed *rafter* roof that has largely replaced it.

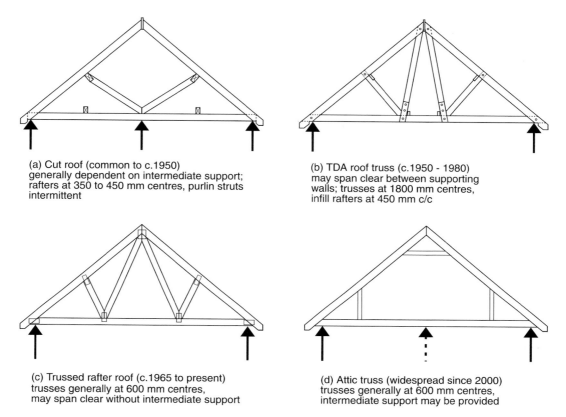

(a) Cut roof (common to c.1950)
generally dependent on intermediate support;
rafters at 350 to 450 mm centres, purlin struts
intermittent

(b) TDA roof truss (c.1950 - 1980)
may span clear between supporting
walls; trusses at 1800 mm centres,
infill rafters at 450 mm c/c

(c) Trussed rafter roof (c.1965 to present)
trusses generally at 600 mm centres,
may span clear without intermediate support

(d) Attic truss (widespread since 2000)
trusses generally at 600 mm centres,
intermediate support may be provided

Fig. 9.1 Domestic roofs: predominant structural forms.

The TDA truss was developed by the Timber Development Association, now TRADA, as a response to restrictions on the use of timber that followed World War II, and it was first used in 1947. While its primary purpose was to reduce the amount of timber used in roof construction, it also allowed for the creation of greater clear spans in dwellings than had been typical before 1939. For this reason, the presence of internal load-bearing walls should not be assumed.

In a TDA truss, the primary triangular frame is created by two pairs of opposing rafters linked by a horizontal ceiling tie (Fig. 9.1b). Additional bracing within the truss is provided by struts, hangers and inclined ties. All components in the truss are fastened by bolts with split ring or toothed plate shear connectors to create a substantial and highly rigid structural frame.

The TDA truss acts as a principal, providing intermediate support for other structural members in the roof. Trusses are generally fixed at 1800 mm centres and support purlins and ceiling binders; these, in turn, support common rafters and ceiling joists set between them at 450 mm centres. In this sense, it is still very much a traditional ridge and purlin roof structure, although the design allows for purlins of considerably reduced section. A number of standard designs for bolted trusses were developed by the TDA and later TRADA.

Some elements of the TDA roof configuration are lighter in section than those of comparable components in pre-war roofs, and the spacing between them is slightly greater (450 mm rather than 400 mm for rafters and joists). Nevertheless, these roofs are generally more generously proportioned than the trussed rafter roofs that have replaced them. Note that in contemporary trussed rafter configurations, spacing is generally 600 mm.

Trussed rafter roofs (1965 to present)

There is a perception that trussed rafter roofs cannot be converted. This is incorrect. Timber trussed rafter roofs are certainly neither the easiest to convert nor the cheapest, but conversion is generally feasible provided there is adequate headroom.

In the conventional trussed rafter roof, trusses are generally fixed at 600 mm centres (Fig. 9.1c). Structural webbing and stability bracing occupy a substantial portion of the void and this makes working in the roof awkward. The major components, such as rafters and chords, are quite modestly proportioned compared to equivalent elements in cut roofs, and even minor alterations must be justified by calculation. In addition, the emergence of the trussed rafter form coincided with the vogue for much lower roof pitches and consequently reduced headroom. Note that trussed rafter roofs are often intended to span between wall plates without intermediate support. The presence of internal load-bearing walls should not be assumed.

The attic or 'room-in-roof' (RiR) truss is a variant form of the trussed rafter and is now widely used in new build (Fig. 9.1d). Webbing in the conventional sense is largely absent and a roof formed with such trusses therefore provides a useable void. Although the RiR truss emerged more than 30 years ago, it has only become popular in new build since the late 1990s. Notes on attic trusses are provided in a separate section at the end of this chapter.

CUT ROOF: STRUCTURAL FORMS

The cut roof encompasses a considerable number of structural configurations but only common domestic forms are considered here. Generally, a distinction is made between the *single roof*, which has no purlins and the *double roof*, where purlins are present. Note that a dwelling may contain both single and double roof elements. For example, a terraced house might have a principal double roof; the back addition (see Glossary) of such a dwelling would generally have a 'lean-to' (single) roof. It should be noted that roofs do not always occur in 'pure' forms.

Single roofs

These are constructed with rafters supported at their head and feet only: no transverse intermediate support is provided by purlins, although collars and ties are present in some of the versions described below. The span of such single roofs is relatively limited. Basic forms of the single roof include:

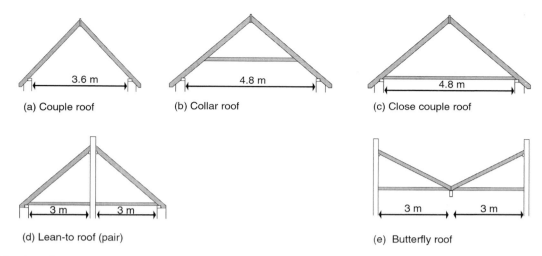

Fig. 9.2   Common single roof configurations (figures indicate typical maximum spans).

- *Couple roof*: rafters are pitched from the wall plate to the ridge (Fig. 9.2a). No horizontal tying elements are provided. Maximum span on plan: 3.6 m. Not widespread as a domestic form.
- *Collar roof*: rafters are pitched from wall plate to ridge (Fig. 9.2b). Horizontal collars are fixed between rafter pairs about one third of the way up the roof slope; rooms formed beneath the collars are thus partly within the roof void. Maximum span on plan: 4.8 m.
- *Close couple roof*: rafters are pitched from wall plate to the ridge (Fig. 9.2c). Horizontal ties are provided at rafter foot level to limit the outward thrust of the roof slopes. Maximum span on plan: 4.8 m.
- *Lean-to roof*: a mono-pitch roof with rafters running from a wall plate at the foot to wall plate or bearing plate at the top (Fig. 9.2d). Common method of roofing rearward subsidiary projections of terraced dwellings (back additions) where the roof slope of the projection is at right angles to the principal roof. The rafter head is fixed to a party wall. Ceiling joists are generally present, but may not form a true tie. Maximum span on plan: 3 m. Roofs of this sort were sometimes constructed with purlins and ceiling binders to offer a greater span.
- *Butterfly roof*: sometimes called a 'London' or V-roof (Fig. 9.2e). Rafters are pitched to opposing compartment walls from a central timber beam. Generally associated with terraced dwellings particularly in central and inner London (c. 1790 to 1870). Ceiling joists are present, but generally do not provide a true tie. Often characterised by an exceptionally shallow roof pitch (to 15 degrees). Maximum span on plan: 3 m + 3 m. As with the lean-to roof, a butterfly roof could be constructed with purlins to provide a greater span.

Double roofs

Rafter length (clear span) is generally the limiting factor in the construction of the single roof. In a double roof, however, considerably greater spans are achieved by

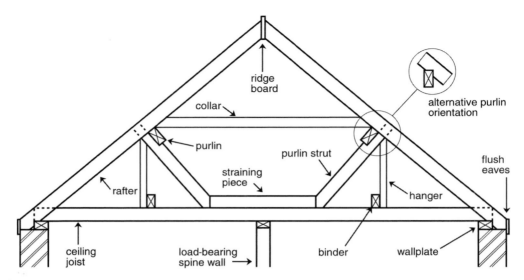

Fig. 9.3   Common double roof elements.

providing intermediate support for rafters by introducing purlins (Fig. 9.3). Double roofs are far more common in dwellings suitable for conversion, and a number of external forms are represented by this mode of construction, including gabled and hipped roofs. Equally, there are a considerable number of internal structural forms. Some of the more common configurations are illustrated in Fig. 9.4.

CUT ROOF: STRUCTURAL ELEMENTS

The following notes provide a description of the functions of the major components in the both gabled and hipped cut roofs (Fig. 9.5). Modifications to elements of the roof structure are outlined in *Common conversion alterations* below. Note that spacing between elements, such as rafters and ceiling joists, is not always regular.

Purlin

The purlin is generally the most substantial of the timber members in the traditional cut roof. They are fixed in a horizontal position and provide intermediate support to reduce the span and therefore the section size of the rafters. Depending on the orientation of the purlin (Fig. 9.3), rafters are fixed by skew nailing directly to the purlin face, or are nailed and notched to the purlin with a bird's mouth cut to the rafter.

In slopes between 4 and 5 m in length, support is generally provided by a single purlin halfway up the slope. To limit rafter span in dwellings with a larger plan area and longer slopes, purlins may be positioned at one third and two thirds of the way along the slope.

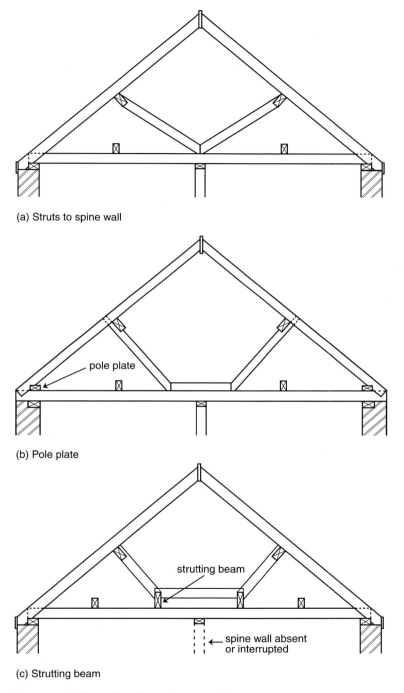

(a) Struts to spine wall

(b) Pole plate

(c) Strutting beam

Fig. 9.4 Common double roof configurations – variations.

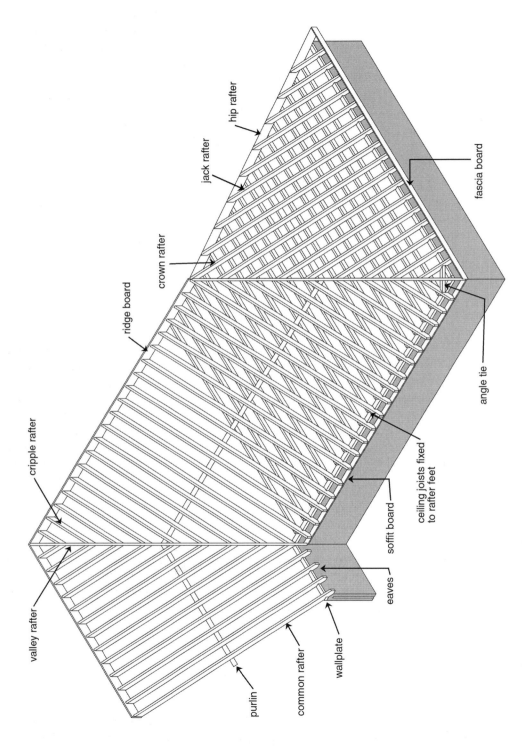

Fig. 9.5   Cut roof: basic structural elements.

In a single purlin slope, the effect of removing the purlin is to double the span of the rafters and, in most cases, this would not be acceptable unless an alternative means of support were provided for the slope. Table 9.1 (p. 189) provides an indication of maximum spans for rafters with dimensions of 100×50 mm (the nearest metric equivalent to the imperial $4" \times 2"$ section).

Purlins generally require intermediate support and this is provided in a number of ways depending on the design of the roof and the building. Note that the ends of purlins are often found to form a short cantilever, particularly in hipped roofs. Support for purlins may be provided in the following ways:

Gable wall support for purlin

A purlin is sometimes built into a supporting masonry gable although, for reasons of fire safety, there is a longstanding presumption against this. Support may also be provided by brick corbels projecting from a masonry gable (common, see Fig. 5.4) or by a shoe or hanger fixed to the gable. In many cases, the purlin abuts the gable wall but support is provided by struts rather than the wall itself.

Strut support for purlin

This is a common method of support for purlins. In a typical pitched roof configuration, inclined purlin struts for opposing slopes are positioned directly opposite each other with the load from the purlins transmitted to a central spine wall. Alternatively, in cases where a central spine wall is not present, the struts may be supported by binders at ceiling joist level (sometimes called strutting beams) that are suitably dimensioned for both roof and ceiling loads (Fig. 9.4c).

Typically, purlin struts are positioned at 90 degrees to the roof slope, generally in opposing pairs. A horizontal straining piece may be introduced at the base of the struts on wider spanning roofs (Fig. 9.3). Purlin struts should not be confused with binder hangers which are generally fixed vertically near the purlin and are supported by the roof structure.

Principal rafter support for purlin

In this configuration, intermediate support for the purlins is provided by a truss within the roof structure itself, rather than by an internal wall. Support may be provided by a traditional principal rafter truss (generally a pair of tied rafters of substantially greater section than the common rafters) or a TDA-type truss described at the beginning of this chapter (Fig. 9.1b).

Ridge and rafters

Ridge or ridge board

This is a horizontal timber member at the apex of the roof to which the heads of the common rafters are nailed. Typically, this is between 25 mm and 32 mm in breadth. The ridge is generally configured to project 25 mm above the rafter heads.

Common rafters

These are pitched from the wall plate to the ridge and provide support for the roof covering, usually via battens. There is generally a plumb cut to the head of the rafter (rather than a bird's mouth) and the rafter is fixed by nailing to the ridge board. In order to achieve balance, rafters of opposing slopes are fixed directly opposite each other at the ridge. The foot of the rafter is generally fixed to the outer edge of the wall plate with a bird's mouth notch and skew nailing. In all cases, the bird's mouth must not exceed one-third of the rafter depth. Note that the rafter feet may be bird's mouthed to the *inside* of the wall plate where eaves are sprocketed. Where a purlin provides intermediate rafter support, the rafter is nailed and is sometimes bird's mouthed over it as well. Rafters are not necessarily continuous over purlins. In older buildings, rafter spacing is not always equal across the slope.

Valley rafter (valley structures)

This provides a fixing for cripple rafters where roof slopes meet to form an internal angle. Note that the upper edge of the valley rafter does not project above the cripple rafters.

Hip rafter (hip structures)

The inverse of a valley rafter. It provides a common fixing for the heads of jack rafters in a hipped roof configuration. The hip rafter is considerably deeper in section than the rafters it supports. The upper edge of the hip rafter does not project above the jack rafters.

Hip board (hip structures)

Sometimes used as an alternative expression for hip rafter (above). It also has a different meaning and is used to describe boards fixed either side of the hip rafter to provide support for slates.

Crown rafter (hip structures)

The central rafter in a run of jack rafters in a hipped configuration. It forms a junction with both hips at the apex. Occasionally omitted.

Jack rafters (hip structures)

These run from the wall plate to the hip rafter with a compound angle cut to the head. As with common rafters, they are skew nailed and fixed in pairs in order to maintain balance.

Cripple rafters (valley structures)

Run from the ridge to the valley rafter. They are skew nailed in opposing pairs. Also described as jack rafters.

Saddle board (hip structures)

A board fixed vertically to the last pair of common rafters at the end of the ridge. It provides a fixing for crown and hip rafters.

Wall plates

These horizontal timber members are fixed to the head of the wall to provide a bearing for rafter feet and ceiling joists. In both solid masonry and cavity masonry structures, wall plates are usually set in line with the inner face of the wall. Wall plates are generally 100×75 mm or 100×50 mm. They are bedded level in mortar to provide an even bearing surface. At junctions and breaks, it is practice to form half-lapped joints.

Strapping for wall plates and rafters

Traditionally, vertical restraint fixings were not used for wall plates and rafters, the mass of the roof structure and its tile or slate covering being considered sufficient to resist both lateral and uplift forces generated by wind loading; occasionally, plates were nailed through into the brickwork.

Current guidance for new roof structures contained in section 2C of Approved Document A indicates that vertical strapping at least 1 m in length should be provided at eaves level at intervals not exceeding 2 m. Such strapping may be omitted if the roof:

- has a pitch of 15 degrees or more, and
- is tiled or slated, and
- is of a type known by local experience to be resistant to wind gusts, and
- has main timber members spanning onto the supported wall at not more than 1.2 m centres

Note also that there may be a need to retro-fix vertical restraint straps to an existing pitched roof slope if the loading conditions are changed.

Ceiling joists and collars

Ceiling joists provide the primary ties between opposing roof slopes in the majority of dwellings. The purpose of these ties is to limit the tendency of a pitched roof to push outwards against (spread) its supporting walls. They are fixed to wall plates and rafter feet at both sides of the building. Ceiling joists also provide support for ceilings. Because the spans involved are often considerable, each tie may comprise two ceiling joists nailed together at a central point, in effect creating a single continuous member.

Collars are sometimes provided in addition to ceiling joists to enhance the tie between slopes. Note that the collar roof (Fig. 9.2b) is a structural form in its own right in which the collar performs the primary tying function.

CUT ROOF: COMMON CONVERSION ALTERATIONS

Modification of the roof structure

In order to create a useable roof void free from structural intrusions, the system of support for the original roof slope or slopes will be substantially altered (Figs. 9.6 and 9.7). The following modifications may be carried out:

- Removal of original purlin(s) and its ancillary struts. Replacement support for rafters and roof slope may be provided by a purlin wall or steel section instead.
- Rafter heads may be fixed to a new roof beam positioned as closely as possible to apex of the roof, or at a point that provides reasonable headroom.
- Rafters may be fixed to the new floor structure if low-level ties (such as original ceiling joists) are removed. In most cases, the bird's mouthed feet of the rafters will remain fixed to the original wall plate and are not disturbed, particularly on front roof slopes.

Reasons to remove a purlin

Existing purlins frequently present an obstacle to the loft conversion process in traditional cut and TDA truss roofs, particularly so in small and medium-sized dwellings with a single purlin for each slope.

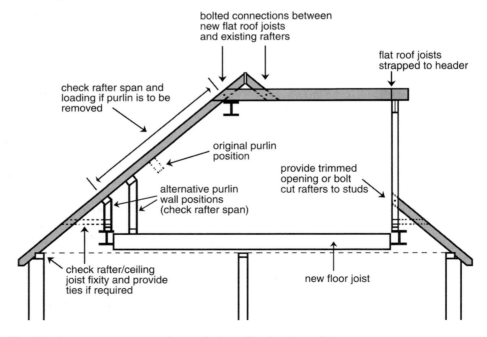

Fig. 9.6 Large dormer conversion: typical modifications to roof structure.

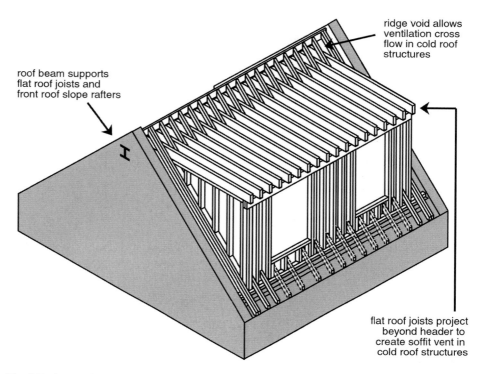

roof beam supports
flat roof joists and
front roof slope rafters

ridge void allows
ventilation cross
flow in cold roof
structures

flat roof joists project
beyond header to
create soffit vent in
cold roof structures

Fig. 9.7 Large dormer conversion: flat roof.

- The purlin in its existing position fouls dormer construction or limits headroom on the new stair. On a front (i.e. highway-facing) roof slope, purlin height is generally critical in relation to the requirement to provide a means of escape: in many conversions, the original purlin would conflict with an escape window (see Chapter 4 for escape window positions).
- The purlin depends on support by strutting or a principal rafter configuration, but these must be removed to increase the useable void as part of the conversion.

Replacement support for purlins

The purlin's role is to reduce the span of rafters, while the role of the struts or principals is to reduce the span of the purlin. When a purlin and its struts must be removed, some other way must be found of providing support for the roof slope.

Replacement support is generally provided by a purlin wall (Fig. 9.6). A timber plate is fixed to the new floor joists (or shot-fired to the floor beam) to support timber studs that are, in turn, fixed either to a bearing plate nailed to the underside of the rafters or directly to the existing rafters by bolting. When adopting this approach, a vertical stud is generally provided for each rafter on the slope.

The terms dwarf stud wall and knee wall are also used to describe the new supporting structure. The vertical supporting timber members are sometimes called ashlar studs, although the term should be used with care because ashlar (or ashlaring, sometimes ashlering) is also widely used to describe non-structural eaves infill.

It is practice to provide access doors to the void behind the purlin wall to allow for storage in the eaves. Note that floor decking is generally laid directly up to the eaves in order to satisfy fire and sound resistance requirements. In order to simplify the ventilation of the roof structure, thermal insulation material is fixed between the studs of the purlin wall, rather than between lower ends of the rafters where it might interfere with eaves ventilation. Access doors to the void must also be insulated.

The critical factor in the positioning of the purlin wall is its relationship to the roof slope it must support. This is determined by the span and spacing of the existing rafters, their sectional dimensions and strength, and roof loading conditions. A preliminary indication of the maximum possible upper slope span may be obtained by reference to section 4.3 of the TRADA *Span tables*. For indicative purposes, Table 9.1 (p. 189) indicates typical maximum spans for a common rafter section size. Note also that the supporting floor structure or supporting beam must be configured to accept a roof loading.

For the practical purpose of providing structural support for the roof slope *before* the purlin is removed, the purlin wall is generally fixed somewhat closer to the eaves than the existing purlin line subject to structural engineering calculations. This has the advantage of providing a greater floor area for the conversion, but there is little practical advantage obtained by a height of less than 600 mm. Note that there are no minimum headroom requirements for rooms.

Support for the roof slope, whether temporary or permanent (e.g. a purlin wall), must always be provided *before* an existing purlin and any of its supporting structure are removed (Fig. 9.8). Apart from the risk of structural failure, it is

Fig. 9.8 New purlin wall with remnant of original purlin still in situ.

important to realise that even small movements of the existing roof structure caused by alteration are almost impossible to recover once they have occurred.

Roof beam support for rafters

Typically, a timber ridge board of the sort described earlier is generally intended only to locate rafter pairs in an existing roof. It provides a limited degree of restraint, and functions in a satisfactory manner only when balancing the rafters of opposing slopes.

In larger dormer conversions (generally more than 1.5 m width), or where a substantial part of one roof slope is removed, a roof beam spanning from gable to gable must be provided. Typically, this provides a bearing for rafters in the existing roof slope and supports the new dormer's flat roof joists (Fig. 9.9). Depending on the design of the conversion, the beam may be fixed in a number of positions at or near the apex of the roof (Fig. 9.10).

A universal beam or column is commonly used in this application. A flitch beam may also be used at the ridge provided it is adequately restrained against lateral movement.

Timbers fitted into the webbing and bolted or shot-fired through the neutral axis of the beam provide a bearing surface for fixing. Typically, webbing timbers are bolted at 600 mm centres, although it is advisable to ensure that bolt projections do not coincide with fixing points for rafters and joists. Shot-fired fastenings fixed with a powder-actuated tool are increasingly used in this application.

Fig. 9.9 Roof beam configuration. Beam infill timbers provide a bearing for rafters (left). Flat roof joists (right) are notched into the web – note slight downstand to allow for shrinkage.

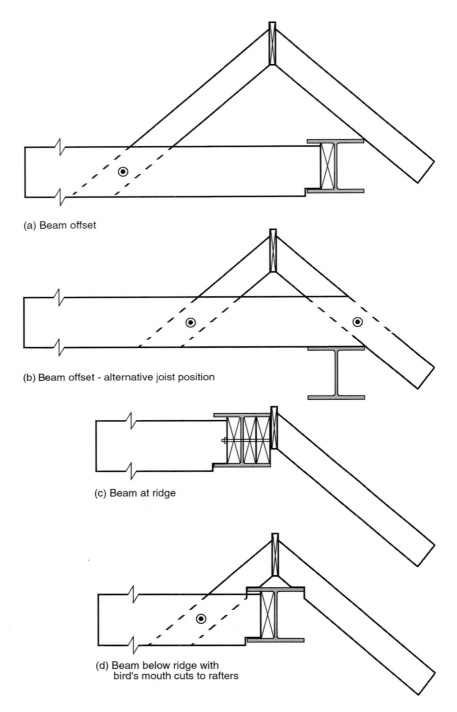

(a) Beam offset

(b) Beam offset - alternative joist position

(c) Beam at ridge

(d) Beam below ridge with
    bird's mouth cuts to rafters

Fig. 9.10 Large dormer conversion: roof beam positions.

RAFTERS

Trimming

Trimming members are required to provide reinforcement for the roof structure when rafters are removed to create an opening (Fig. 9.11). The techniques used to create trimmed openings in roof slopes are broadly similar to those described for trimming floor joists (Chapter 7), but there are variations in framing depending on the type of opening being created. There are three elements common to framing any trimmed roof opening:

- *Trimming rafters* are generally pitched from the wall plate to the ridge, with intermediate purlin or purlin wall support.
- *Trimmed rafters* are the existing common rafters that are cut to accommodate the roof opening.
- *Trimmers* run at right angles to the roof slope and support the cut ends of the trimmed rafters at the head and foot on the opening. Trimmers must be positively fixed to the trimming rafters.

Trimming for roof windows

Structural openings for proprietary roof windows are generally relatively straightforward to create in traditional cut roofs. Note that double and treble

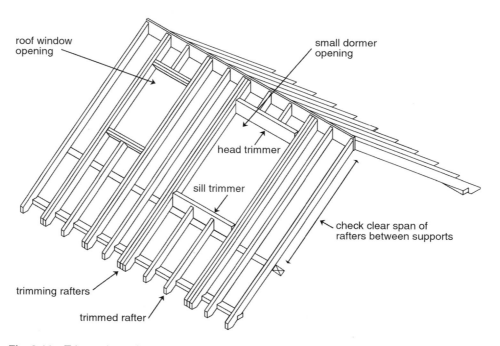

Fig. 9.11 Trimmed openings.

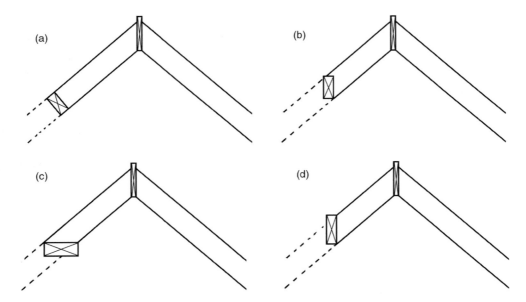

Fig. 9.12   Head trimmer detail.

trimming rafters may be required for larger windows. Double trimmers may be required to support trimmed rafters at the top and bottom of the opening. Historically, trimmers were fixed to the trimming rafters with pinned tenon joints, with the trimmed rafters dovetailed to the trimmer. In modern practice, timber elements are generally butted and nailed together (Fig. 9.12a). Multiple trimming rafters are bolted together.

Trimming for dormer projections (general)

The principles for framing a trimmed opening for a dormer projection are similar to those outlined above but there are two key differences. The load of a dormer projection is likely to be considerably greater than that of a roof window occupying a similar area of roof slope, and careful consideration must be given to sizing the trimming rafters. In addition, the relationship between the trimmers and the dormer structure must also be considered in order to simplify construction.

Small dormer windows

These generally occupy only part of the roof slope. Rafters are trimmed to accommodate the projection. The trimmer at the top of the opening is generally described as a head trimmer and this may be oriented in a number of ways depending on the design of the dormer (Fig. 9.12b–d).

In a small flat roof dormer, the head trimmer may provide direct support for both trimmed rafters and the dormer's flat roof joists. It is therefore configured in a vertical position relative to its axis to facilitate fixing of the dormer roof joists.

The trimmer at the foot of the opening is generally described as the sill trimmer and, like the head trimmer, it may be fixed in a vertical position relative to its axis where a dormer is to be accommodated. It is also generally configured to project above the plane of the rafters to facilitate the fixing of roofing materials and flashings. Historically, trimmed rafters at the head and sill were bird's mouthed over the trimmer; in current practice, they are generally plumb cut and nailed.

The dimensions and numbers of trimming members required for small dormers are determined by several factors. These include the width and length of the opening formed in the roof structure, the new load and the structural support available (for example, support by purlins or floor structure, see Fig. 9.13). Note

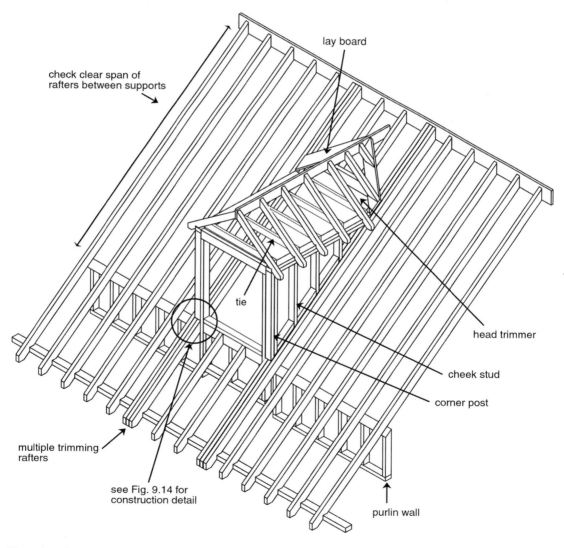

Fig. 9.13   Small dormer: construction and trimming.

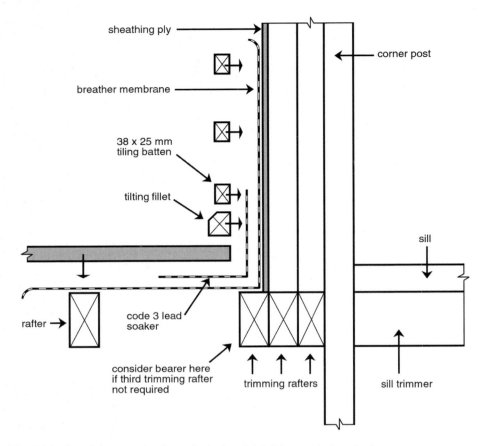

sheathing ply

breather membrane

38 x 25 mm
tiling batten

tilting fillet

corner post

sill

rafter

code 3 lead
soaker

consider bearer here
if third trimming rafter
not required

trimming rafters

sill trimmer

Fig. 9.14   Small dormer: cheek to principal roof detail (expanded section).

that, in larger traditionally constructed dwellings the roof may be configured with two purlins. These are generally set at one-third and two-thirds of rafter span, typically on slopes of up to 6 m. In some cases, these are configured to act as trimmers in their own right for original dormer structures.

In many cases, the need to provide a fixing for cheek studs and roof battens to the principal slope means that it may be prudent to provide additional trimming rafters even if there is no prima facie structural engineering case for doing so (Fig. 9.14). Note that in order to accommodate appropriate levels of insulation, it is generally necessary to frame the walls and roofs of small dormer projections in 100×47 mm studwork (Figs 9.15 and 9.16).

Large box dormers

In cases where a full-width dormer conversion occupies the whole of a pre-existing rear roof slope there will clearly be no need for trimming because the entire roof slope is removed and the remaining front slope is supported by a roof

Fig. 9.15 Hipped dormer. Note the provision of a low-level tie.

Fig. 9.16 Hipped dormer detail with rigid PIR insulation between studs.

beam or beams. However, many large box dormers are subordinate structures to some degree, and portions of the roof slope above, below and to the sides of the dormer are retained and must be provided with support. Where a dormer exceeds 1.5 m in width, it is generally necessary to support both the dormer roof and the principal roof with a beam spanning between gables: a trimmed opening alone would usually not be adequate.

- *Lower portion of original roof slope (eaves to base of dormer)*: trimmed rafters may either be fixed to a trimmer which is usually vertically oriented on its axis, or fixed directly to the new box dormer face studwork by bolted connections.
- *Upper portion of original roof slope (back of dormer flat roof to existing ridge)*: where this part of the slope is to be retained, the trimmed rafters are fixed directly to new flat roof joists, generally by bolting, thereby eliminating the need for an independent trimmer. Note that the new flat roof joists must be set at centres to match the existing rafter spacing. This approach is adopted where one of the primary supports for the new flat roof joists is provided by a load-bearing roof beam (Fig. 9.10a, b and d).
- *Side slopes of original roof (eaves to ridge)*: many box dormers are subordinate structures only to the extent that they are formed immediately within the gable walls of the dwelling. The margin of remaining roof slope flanking the dormer may thus be supported by the gable and in such cases there is no rafter to trim. Where the dormer is narrower, and original rafters retained on one or both sides, these may be reinforced where necessary by the introduction of supplementary trimming rafters. Additional support may also be provided by cheek studwork, where this is supported by a structural floor, or by the provision of a purlin wall.

Sizing and loading of rafters

The complexity of alterations carried out during loft conversions means that it is usually not possible to specify rafters and rafter spacing using span tables. The introduction of trimming loads, repositioning of purlins and the additional loading brought about by the introduction of insulating material and plasterboard finishes to the inside of slopes means that, in most cases, justification for rafter spacing must be made by calculation.

Where new rafters are introduced to reinforce a roof slope, they must generally be of the same section size as the existing rafters in order that they may be accommodated within the thickness of construction. It should be noted that rafter spacing is often not equal in older buildings and, in addition to this, span tables do not take spacings closer than 400 mm into account. Table 9.1 is provided for indicative purposes only.

Table 9.1 50×100 mm rafter – maximum clear spans (roof pitch 30 to 45 degrees).

Joist spacing (mm)	C16 Clear span (m)	C24 Clear span (m)
400	2.38	2.49
450	2.30	2.40
600	2.09	2.18

Imposed snow loading 0.75 kN/m²
Access only for maintenance or repair
Dead load 0.75 to 1.25 kN/m² (excluding self weight of rafter)
Based on Approved Document A (1992). TRADA's span tables (2005) do not reference 50×100 mm rafter sections.

HIP-TO-GABLE CONVERSION

Hip-to-gable conversions are carried out in order to increase the volume and useable full-headroom floor area in a loft conversion. An existing section of hipped roof, generally to the side of the building, is removed, and the original roof extended to meet a new gable wall. The new gable may be constructed in masonry or timber studwork. Note that a correctly configured studwork gable wall is capable of providing support for steel ridge and floor beams.

In order to facilitate the introduction of new rafters, the original hip rafters are removed. On the slopes that are retained (generally the front and sometimes the rear roof slopes), the jack rafters and the roofing material covering them are left in position: as far as possible, the existing slopes are left undisturbed. New rafters are fixed, generally by bolting, beside the existing jack rafters. The new longer rafters are generally pitched from the level of the new purlin wall to the new ridge plate or new roof beam. Slates or tiles recovered from the redundant hip may provide a source of well-matched roofing material for the newly extended slope or slopes.

As with any structural alteration, careful attention should be given to the provision of both temporary and permanent supports whilst work is being carried out.

NOTCHES AND HOLES

It may be necessary to provide structural engineering calculations to justify cutting holes or notches into elements of roof structure, for example, to permit continuous ventilation around roof windows (Fig. 9.20). Note that only floor and flat roof joists may be drilled or notched without formal calculation, as long as guidance is followed: this is reproduced in Chapter 7.

LATERAL SUPPORT FOR GABLES

In order to resist forces acting horizontally (such as wind loading), guidance in Approved Document A *Structure* indicates that gable walls should be strapped to the roof structure. When a loft is converted, the extent of compliance with this guidance depends on the nature of the alterations to the roof. In most traditionally constructed cut roofs, restraint fixings were not provided and, in some cases, it may be prudent to fix them as part of the conversion even if it is not a requirement.

- *Existing pitched roof slope unaltered.* There may not be a requirement to retro-fix lateral restraint strapping if a roof slope is substantially unaltered by building work, for example, in the case of a front roof slope to a gable-ended dwelling.
- *Hip-to-gable conversion.* Where an entirely new gable is built, either in masonry or studwork, to replace a roof hip, it is necessary to provide lateral tension straps between the new wall and the new roof structure. This applies equally to the construction of a flank gable wall (see also Chapter 8).
- *Provision of lateral support for gable walls.* A method for providing lateral support for gable walls is described in Approved Document A, section 2C. Tension straps (generally galvanised mild steel 30×5 mm) are fixed between the roof slope and the gable wall. A strap is provided at or near the highest point in the roof, and additional straps fixed at no more than 2 m centres down the roof slopes.

Note that, depending on the ratio of gable wall thickness to height, it may also be necessary to provide restraint fixings between the new floor structure and gable wall in some cases.

REPLACEMENT ROOF COVERINGS

In cases where it is proposed to replace the roofing material on an existing roof with one of significantly greater mass, the structural integrity of both the roof and its supporting structure must be checked. Approved Document A *Structure* indicates that 'significant' in this context means an increase in roof loading of 15% or more.

For example, replacing slates (unit load approximately 0.30 kN/m^2) with concrete tiles (unit load approximately 0.51 kN/m^2) would result in a 'significant' change in load. Note that any strengthening work or replacement of roof members is classified as a material alteration.

Also, where a replacement roofing material has a *lower* mass than the original, the roof structure and its anchoring should be checked to ensure that an adequate factor of safety is maintained against the risk of wind uplift.

FLAT ROOF: BASIC STRUCTURE

Flat roofs are an element common to all box dormer conversions. Typically, flat roof joists are fixed to a roof beam shared with the existing pitched roof at the ridge (Fig. 9.10) and are strapped to the header at the dormer face (Fig. 9.17). The slope of the flat roof is generally provided by tapered timber firring pieces fixed to the top of the roof joists and is laid to fall to the rear of the building (i.e. towards the dormer face) where guttering must be provided. Flat roof joists may be selected from TRADA's *Span tables*. An indication of permissible clear spans is

Fig. 9.17 Box dormer. Sheathing and roof deck in 15 mm OSB. Flat roof joists are strapped to the header; inverted 'long-leg' hangers have been used in this example.

Table 9.2 Flat roof joists (C16) – permissible clear spans.

Joist size (mm)	Joist spacing (mm)	Clear span (m)
47 × 147	400	2.96
47 × 170	400	3.53
47 × 195	400	4.14
47 × 220	400	4.75
75 × 147	400	3.56
75 × 170	400	4.21
75 × 220	400	5.50

Maintenance/repair access only
Imposed snow load 0.75 kN/m^2
Dead load 0.75 to 1.00 kN/m^2 (excluding self weight of joist)
For general guidance only. Please refer to TRADA span tables (2005)
Source: Approved Document A (1992)

given in Tables 9.2 and 9.3. These are based on the span tables included in the 1992 version of Approved Document A – note that TRADA's 2005 tables generally give slightly greater spans.

Flat roofs take two forms. Both are similar in terms of their basic structure but differ in the ways that they are insulated and in the requirement to provide ventilation. Note that strutting should be provided between the flat roof joists – see Chapter 7, *Strutting*.

Table 9.3   Flat roof joists (C24) – permissible clear spans.

Joist size (mm)	Joist spacing (mm)	Clear span (m)
47 × 147	400	3.11
47 × 170	400	3.69
47 × 195	400	4.33
47 × 220	400	4.94
75 × 147	400	3.73
75 × 170	400	4.40
75 × 220	400	5.70

Maintenance/repair access only
Imposed snow load 0.75 kN/m²
Dead load 0.75 to 1.00 kN/m² (excluding self weight of joist)
For general guidance only. Please refer to TRADA span tables (2005)
Source: Approved Document A (1992)

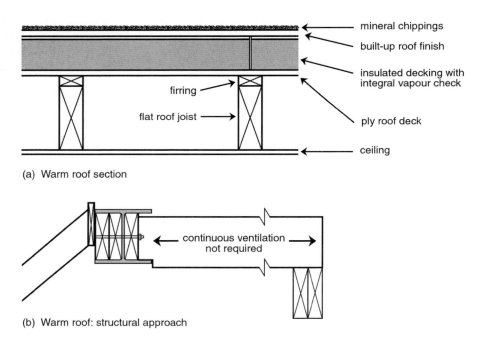

(a) Warm roof section

(b) Warm roof: structural approach

Fig. 9.18   Flat roof: warm deck.

Flat roof – warm deck

In a flat warm roof, the insulating material is carried above the roof joists (Fig. 9.18a). Because the voids between the ceiling finish and the underside of the insulation are normally at more or less the same temperature as the room below them, the risk of moisture-laden air condensing within the structure is reduced and ventilation of the void is therefore not required.

Because the insulation is fixed above the roof joists, a warm flat roof is thicker than a cold one and this is sometimes problematic in conversions where head-room is limited. Note, however, that this system of roof construction offers a number of practical benefits:

- Because there is no need to ventilate the void, a warm roof can be used in applications such as dormer construction where cross ventilation is not always possible.
- Thermal bridging is eliminated and the roof will provide more effective insulation.
- Construction is quicker and less labour intensive.
- Recessed lighting can generally be fixed without the need for fire hoods, and cables need not be derated.

Flat roof – cold deck

Insulation in a cold flat roof is accommodated in the void between the roof joists and sometimes below the joists as well. This results in a roof that has a relatively shallow section, which is advantageous in a conversion where ceiling height is restricted (Fig. 9.19). A cold flat roof, sometimes called a cold deck, requires cross

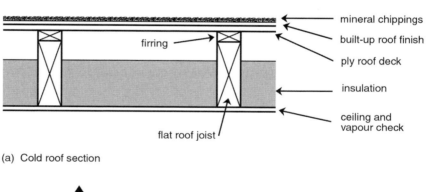

(a) Cold roof section

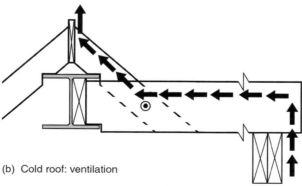

(b) Cold roof: ventilation

Fig. 9.19 Flat roof: cold deck.

ventilation to reduce the risk of water vapour from within the building condensing within or above the insulating material. A vapour check is provided to further minimise this risk. The following factors should be considered:

■ Provision of inward airflow at the eaves – generally at the dormer face by oversailing the flat roof joists to create space for soffit vents. Where the existing ridge structure and part of the original roof slope are retained, it is possible to achieve continuity of airflow to a new or existing ridge vent. In cases where the joists at the ridge end of the roof must be supported by a beam that is fixed at ridge level (Fig. 9.10c), providing cross flow ventilation is not generally possible. Warm deck construction (see above) must be considered in these cases.

■ Because insulation is carried between the roof joists rather than above them, a degree of thermal bridging will occur.

■ The use of recessed lighting creates difficulties where cold roof construction is to be used. ELV (extra-low voltage) lighting systems in particular generate considerable amounts of heat through transformers, cable connections and the lamps themselves, and it is necessary to remove any insulating material surrounding them. In addition, any recessed fittings will breach the ceiling vapour check and allow moisture-laden air from the rooms below into the void. Cables must be derated where they pass through insulation.

CONDENSATION IN ROOF STRUCTURES

Ensuring that a cold roof structure is resistant to damage caused by interstitial condensation is a requirement of the Building Regulations and, to satisfy the requirement, it is generally necessary to provide ventilation. In order to reduce the risk of moisture-laden air from inside a dwelling condensing within the structure (interstitial condensation) a dual strategy is adopted to protect the roofs of dwellings:

■ A vapour check is provided between internal plasterboard and the warm side of the insulation to limit the passage of water vapour from inside the building to the roof structure. This may take the form of vapour-impermeable sheet material such as polythene or wallboards with an integral vapour check.

■ Any moisture-laden air that does enter the roof structure must be allowed to escape. This may be achieved by providing a 50 mm ventilation void between the cold side of the insulating material and the underside of the existing vapour-impermeable roofing felt or roof deck. In pitched roofs where the ceiling follows the roof pitch, airflow through this void is provided by openings equivalent to 25 mm continuous at the eaves and 5 mm continuous at the ridge.

Breather membrane may be provided instead of traditional underlay. Breather membranes allow water vapour to escape but prevent the ingress of liquid water and have the advantage of being easier to handle than traditional underlays.

Pitched cold roofs

A cold roof in the context of a loft conversion is one where the insulating material is carried beneath or between the rafters. A characteristic of most loft conversions is that the ceiling follows the pitch of the roof, at least at some point, and this generally applies to the full length of the slope nearest the highway. To reduce the risk of interstitial condensation occurring in situations such as this, a 50 mm gap is provided between the upper (cold side) of any insulation and the underlay. In tandem with this, ventilation must be provided at the eaves (equivalent to a 25 mm continuous strip) and at the ridge (equivalent to a 5 mm continuous strip).

Providing ventilation at the apex of the roof is generally a relatively straightforward matter of providing proprietary ridge vents. Airflow at the eaves may be achieved through the use of soffit vents. However, in the case of dwellings with flush eaves (Fig. 9.3), providing airflow is sometimes more troublesome. Where soffit or fascia vents cannot be accommodated, proprietary tile or slate vents may be provided. In such cases, it is best that insulation inside the conversion is carried between the studs of the purlin wall (thus creating a continuous cross-void at eaves level), rather than following the roof slope down to the rafter feet. If insulation were to be carried down to the eaves, it would be necessary to provide a vent for each rafter void, which is both costly and unsightly.

Pitched warm roofs

A warm roof is one where the insulating material is carried above the rafters. Because the air within any structural voids is similar in temperature to the room beneath, the risk of condensation occurring is reduced. It is not normally necessary to ventilate warm roof voids.

Pitched warm roofs are a relatively recent innovation. Note that it is seldom practical to provide warm roof insulation to existing pitched roofs in terraced or semi-detached dwellings because it increases the height of the roof envelope relative to the adjacent buildings. In addition, an alteration to the roof of this sort is likely to require planning permission by virtue of Schedule 2, Part 1 (B1) of the GPDO.

Flat roof – cold deck

The ventilation provisions for cold deck flat roofs are similar to those for pitched roofs where the ceiling finish follows the roof pitch (above). A 50 mm gap is provided between the cold side of the insulation and the roof deck, and airflow is provided by the equivalent of continuous 25 mm ventilation at the dormer face and 5 mm at the ridge. Where this cannot be achieved, an alternative is to provide a warm deck roof.

Flat roof – warm deck

Insulation is carried above the joists. As noted earlier, there is relatively little risk of condensation forming within the roof void and it is therefore not necessary to provide ventilation.

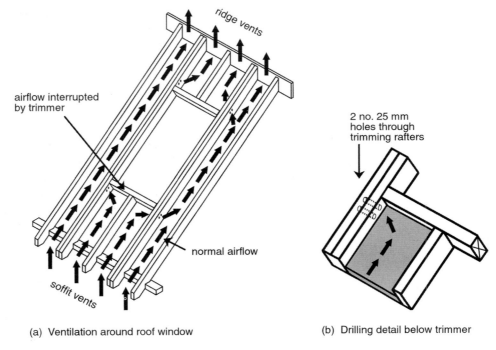

Fig. 9.20 Insulated cold roof slopes: ventilation.

Continuity of airflow around roof windows

In cold pitched roofs, the provision of roof windows will block the ventilation path, and steps must be taken to ensure that adequate airflow is maintained. Where a roof window is to be fixed in an original roof slope, one way of achieving continuous ventilation is to drill 25 mm diameter holes through the trimming rafters immediately above and below the trimmed opening (Fig. 9.20). In wider openings, it is necessary also to drill through the intervening trimmed rafters. To be effective, these holes must be drilled *above* the level of insulating material between the rafters. Note that it would generally be necessary to justify any such proposal by calculation. Alternatively, external vents may be provided immediately above and below the obstruction.

Ventilation

Possible exemptions from the requirement

Section 6.11 of Approved Document C states that:

'For the purposes of health and safety it may not always be necessary to provide ventilation to small roofs such as those over porches and bay windows.'

It is not clear to what extent this guidance would apply to a section of pitched or flat roof on a small dormer projection. It is generally prudent to work on the assumption that *all* cold roofs require ventilation. Any proposal not to ventilate should be the subject of discussion with, or made as part of a full-plans submission to, the building control service chosen.

Approved Document guidance

Guidance on protecting roofs from condensation was originally included in Approved Document F2. In 2004, however, this guidance was moved to Approved Document C2. Advice on methods of ventilating roof voids has been removed and, instead, Approved Document C indicates that the requirement to resist damage from interstitial condensation will be met by designing a roof in accordance with clause 8.4 of BS 5250:2002 and BS EN ISO 13788:2001, with further guidance provided in BRE Report BR 262.

ATTIC TRUSSES

The trussed rafter roof was first demonstrated in Britain in 1963 and was rapidly adopted for domestic roof construction during the 1970s. It remains the dominant method of domestic roof construction in the UK (Fig. 9.21a).

However, pressure to increase housing density has forced housebuilders upwards and has led to a radical reappraisal of the way roof voids are used. Since 2000, production of attic or room-in-roof (RiR) trusses has increased five-fold and RiR trusses now account for almost a third of trussed rafter production. Attic trusses are configured to yield a useable roof void and are designed for domestic floor loading (Table 9.4). These are increasingly being installed as standard in new build (Fig. 9.21b).

In certain circumstances, it is possible to use attic trusses to replace completely the roof structure of an existing building. There are two principal approaches. One is to remove the existing roof and ceiling structure in a single operation, although clearly removing the ceiling is a highly disruptive procedure. The other method is to replace existing trusses with new attic trusses one at a time, tying the existing bottom chords (which carry the ceiling) to the new trusses as work progresses.

The main drawbacks with a total roof replacement are likely to be high crane handling costs, and the need to provide temporary storage for the new trusses

Table 9.4   Room-in-roof trusses – minimum practical dimensions.

Roof pitch	Minimum span	Room width between supporting walls
35 degrees	9 m	4.5 m
40 degrees	7 m	4.0 m
45 degrees	6 m	3.5 m

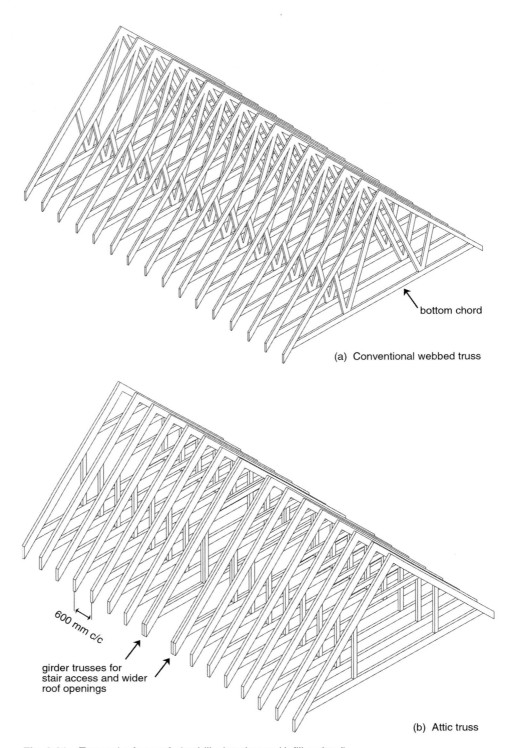

(a) Conventional webbed truss

bottom chord

600 mm c/c

girder trusses for
stair access and wider
roof openings

(b) Attic truss

Fig. 9.21   Trussed rafter roofs (stability bracing and infill omitted).

when they are delivered. Most terraced houses, for example, are deeper than they are wide, and even a reasonably-sized front garden is unlikely to be able to accommodate the trusses required to convert the roof.

Note also that the arrangement of sole plates and wall plates in the roof to be converted must be perfectly regular in order to accept the new trusses. Dimensional regularity in older masonry-constructed dwellings should not be assumed and trusses are generally intolerant of even minor modifications.

10 Energy conservation

The requirement to conserve fuel and power is set out in Part L of the Building Regulations. The primary purpose of Part L is to save carbon and therefore to limit carbon dioxide emissions created both directly and indirectly by buildings. Guidance on complying with Part L is provided in Approved Document L *Conservation of fuel and power*.

A new version of Part L with accompanying Approved Document guidance came into effect on 6 April 2006. Approved Document L (2006) is arranged in four parts. Guidance relevant to loft conversions is contained in section L1B (ADL1B), which deals specifically with work in existing dwellings.

Approved Document L, section L1B (ADL1B) provides guidance on means of conserving energy whenever building work is carried out in existing dwellings. This is achieved, primarily, by setting standards for windows, lighting, heating and insulation.

Paragraph 2 of ADL1B identifies six types of building work in existing dwellings. These activities include extension, the creation of a new dwelling through material change of use, material alterations, provision of controlled fittings, provision or extension of controlled services and the provision or renovation of thermal elements. With the exception of material change of use, the other five categories potentially apply to loft conversions in single-family dwellings. The nature of the work that is likely to be triggered by these activities is considered in this chapter.

Changes to the Building Regulations (S.I. 2006/652)

A number of changes and amendments have been made. These include a new regulation (4A). Regulation 4A(1) would apply when work to walls, floors and roofs is carried out in an existing conversion (*renovation*, page 207), while regulation 4A(2) is likely to apply where a wall, floor or roof is rebuilt (*replacement*, page 205).

Regulation 9 has been substantially altered. Buildings in conservation areas and listed buildings are exempt from the energy efficiency requirements of the Building Regulations, but only where and to the extent that compliance would unacceptably alter their character or appearance.

WINDOWS AND OTHER OPENINGS

Windows, roof windows, roof lights and doors are relatively poor insulators. In order to limit heat loss through them, the guidance in ADL1B is that openings of

Table 10.1 Reasonable provision when working on controlled fittings (ADL1B, Table 2).

Fitting	(a) Standard for new fittings in extensions	(b) Standard for replacement fittings in an existing dwelling
Window, roof window and roof light	*U*-value = 1.8 W/m²K *OR* Window energy rating = Band D *OR* Centre-pane *U*-value = 1.2 W/m²K	*U*-value = 2.0 W/m²K *OR* Window energy rating = Band E *OR* Centre-pane *U*-value = 1.2 W/m²K
Doors with more than 50% of their internal face area glazed	2.2 W/m²K *OR* Centre-pane *U*-value = 1.2 W/m²K	2.2 W/m²K *OR* Centre-pane *U*-value = 1.2 W/m²K
Other doors	3.0 W/m²K	3.0 W/m²K

Refer to CE66 *Windows for new and existing housing* (EST) for window energy rating data
Source: ADL1B 2006

this sort be restricted to 25% of floor area. To this proportion may be added the area of any windows or doors that, as a result of the work, no longer exist.

However, the guidance states that different approaches may be adopted by agreement with the building control body in order to achieve a satisfactory level of daylighting. The Approved Document refers users to BS 8206 Part 2 *Code of practice for daylighting*.

Window, roof window, roof light and door units should be draught proofed and have an area-weighted average performance that is no worse than that set out in Table 10.1.

The *U*-value or Window Energy Rating of window or roof window fittings can be taken as the value for either (a) the standard configuration as set out in BR 443 *Conventions for U-value calculations* (BRE 2006) OR (b) the particular size and configuration of the actual fitting. In Table 10.1, the *U*-value has been assessed with roof windows in the vertical position.

Where test data and calculated values are not available, reference may be made to Table 6e (*Indicative U-values for windows, doors and roof windows*) in SAP 2005.

FIXED INTERNAL LIGHTING

In addition to conventional fixed lighting, energy-efficient electric lighting should be provided when a loft is converted. Fluorescent light fittings that incorporate lamps with a luminous efficacy greater than 40 lumens per circuit-watt are satisfactory for this purpose (see also definition of *light fitting* in Glossary). Non-integrated assemblies are generally used in this context to ensure that only low-energy fluorescent lamps can be fitted.

Note that lampholders capable of accommodating and running relatively inefficient incandescent lamps (i.e. tungsten and tungsten halogen lamps) cannot be used for the purposes of providing energy-efficient lighting. Bayonet cap, Edison screw and tungsten halogen fittings are therefore not acceptable.

In order to determine the number of energy-efficient light fittings required, it is necessary to consider both the floor area of the conversion and the number of conventional (i.e. incandescent) fixed light fittings that are to be provided. The guidance states that:

'Reasonable provision would be to provide in the areas affected by the building work, fixed energy-efficient light fittings that number not less than the greater of (a) one per 25 m² of dwelling floor area excluding garages or part thereof or (b) one per four fixed light fittings.'

As well as ensuring a minimum provision of energy-efficient lighting, the guidance is designed to discourage overuse of inefficient incandescent lighting. Tables 10.2a and 10.2b (below) provide an illustration of how this works. Table 10.2a ensures that a minimum number of energy-efficient light fittings are installed (irrespective of the conventional lighting provided), while Table 10.2b establishes a proportional relationship between the two systems when a relatively large number of incandescent fittings are used. Both tables should be considered, and the one that gives the highest number of energy-efficient light fittings for a given set of circumstances should be used.

The Approved Document states that the lighting assessment should be based on the extension or the area served by the lighting system as appropriate to the particular case. The guidance does not intend that both energy-efficient and low-efficiency lighting be provided in the same room – i.e. one room with a choice of circuits – because, under this arrangement, there is no guarantee that the energy-efficient lighting would be used. It should be noted that energy-efficient fittings in less-used areas, such as cupboards and storage areas, would not count towards the total.

Table 10.2   Provision of energy-efficient lighting.

(a) Relative to floor area

Floor area of extension (m²)	Minimum number of energy-efficient light fittings required
0–25	1
25–50	2
50–75	3
75–100	4

(b) Relative to conventional fixed light fittings

Conventional fixed light fittings in extension	Minimum number of energy-efficient light fittings required
4–7	1
8–11	2
12–15	3
16–19	4
20–23	5
24–27	6
28–31	7

ADL1B indicates that it may be appropriate to install an energy-efficient fitting in a location that is not part of the building work e.g. on a landing that is not part of a loft conversion. The Approved Document refers readers to GIL20 *Low energy domestic lighting* (EST 2006).

Whilst a formal definition of 'light fitting' is not provided in the context of conventional incandescent lighting, ADL1B states that a light fitting may contain one or more lamps. It would therefore be reasonable to consider a discrete assembly containing a number of lamps (perhaps a chandelier or a spotlight cluster) to be a single light fitting.

However, independent luminaires should be treated as single fittings. For example, it would be reasonable to treat each ELV downlighter in a single circuit as an independent fitting. Extensive use of such fittings would therefore create a concomitant requirement for the provision of compensatory energy-efficient fittings at the ratio indicated in Table 10.2b (i.e. 1:4) and would thus not be practical in many cases. As noted elsewhere in this book, the risks associated with providing recessed lighting in an insulated roof, ceiling or wall structure should be considered. For the purposes of ADL1B, fixed lighting is included in the definition of *fixed building services* (see Glossary).

HEATING AND HOT WATER SYSTEMS

Anecdotal evidence suggests that between a quarter and a third of domestic boilers are replaced when lofts are converted. ADL1B does not require that a boiler be replaced when building work is carried out. However, when an existing boiler is retained, the extended heating system would need to comply with basic provisions for controls – for example, the provision of TRVs (thermostatic radiator valves) where appropriate.

When a heating or hot water system is replaced, a number of actions are triggered. The guidance considers three main areas:

(1) appliance efficiency
(2) the control system for the installation
(3) the insulation of pipes and ducts

In addition, there is an obligation on the person carrying out the work (e.g. a member of an approved Competent Persons scheme) to provide the local authority with a notice confirming that fixed building services have been properly commissioned. The guidance makes a number of references to recommendations set out in the *Domestic Heating Compliance Guide* (NBS 2006). Note that the provision or extension of controlled services (such as heating and hot water systems) is building work and must comply with the Building Regulations.

INSULATION AND THERMAL ELEMENTS

ADL1B uses the expression *thermal element* to describe a wall, a floor or a roof that forms part of the thermal envelope of a building (see Glossary). The guidance

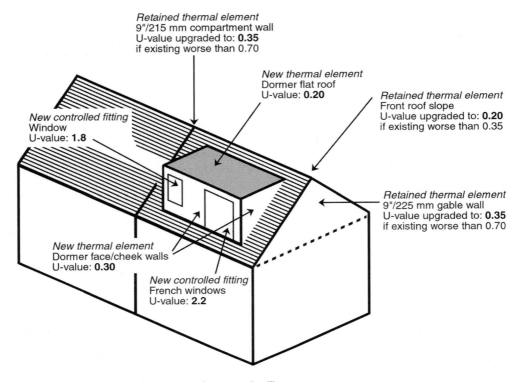

Fig. 10.1   Dormer loft conversion in an end-terrace dwelling.

provided in ADL1B sets out a number of distinct sets of standards for thermal elements:

- Standards for new thermal elements in an extension
- Standards for replacement thermal elements in an existing dwelling
- Standards for retained thermal elements
- Standards for renovation of thermal elements

By its nature, a dormer loft conversion may comprise new elements (such as a new dormer structure), replacement elements (perhaps a new gable wall replaced on a like-for-like basis) and retained elements (such as an existing front roof slope). Therefore in designing the insulation provision for a loft conversion, it would be reasonable to select values from more than one table. Fig. 10.1 illustrates a typical configuration with values drawn from three different sets of standards.

In addition to the main sets of standards, an additional set of figures (*Limiting U-value standards*) is provided. These constitute the minimum acceptable standards in all cases and prevent excessive trade-offs between elements when area-weighted calculations are undertaken. The various standards are described in more detail below:

Table 10.3   Standard for new thermal elements in an extension (ADL1B, Table 4).

Element	U-value (W/m²K)
Wall	0.30
Pitched roof – insulation at rafter level	0.20
Flat roof or roof with integral insulation	0.20
Floors	0.22**

Reference to roof includes roof parts of dormer windows; wall includes dormer cheeks.
In the case of a typical loft conversion, a floor would not normally require thermal insulation. However, it would apply if the roof void over, say, a garage were to be converted.
** See definition below.

Table 10.4   Standard for replacement thermal elements, existing dwelling (ADL1B, Table 4).

Element	U-value (W/m²K)
Wall	0.35*
Pitched roof – insulation at rafter level	0.20
Flat roof or roof with integral insulation	0.25
Floors	0.25**

* A lesser provision may be appropriate where meeting such a standard would result in a reduction of more than 5% in the internal floor area of the room bounded by the wall.
** A lesser provision may be appropriate where meeting such a standard would create significant problems in relation to adjoining floor levels.
Reference to roof includes roof parts of dormer windows; wall includes dormer cheeks.

Standards for new thermal elements in an extension

The meaning of this is reasonably clear in the context of a dormer loft conversion, and would apply to new dormer construction as well as a gable wall constructed as part of a hip-to-gable conversion. Of the three sets of primary standards, those for new elements are the highest (Table 10.3).

Standards for replacement thermal elements in an existing dwelling

In many cases, these would not apply in the context of a loft conversion, unless an existing element (such as a gable wall or roof slope) were to be entirely removed and replaced with an identical new element in the same position. It would also apply if, say, an existing dormer structure were to be replaced on a like-for-like basis (Table 10.4).

Standards for retained thermal elements

These elements, described as 'existing opaque fabric' in the draft Approved Document, would include gable walls and roof slopes (i.e. elements bounding the original roof void) that become part of the conversion's thermal envelope (Table 10.5).

When considering the threshold conditions that are designed to trigger the upgrading of retained thermal elements, the extremely poor performance of

Table 10.5   Upgrading retained thermal elements (ADL1B, Table 5).

Element	(a) Threshold value (W/m²K)	(b) Improved value (W/m²K)
Cavity wall*	0.70	0.55
Other wall type (e.g. 215 mm solid wall)	0.70	0.35
Floor	0.70	0.25
Pitched roof – insulation between rafters	0.35	0.20
Flat roof or roof with integral insulation	0.35	0.25

* This only applies in the case of a wall suitable for the installation of cavity insulation. Where this is not the case it should be treated as for 'other wall type'.

Table 10.6   Existing un-insulated thermal elements – common values.

Element	Unmodified value (W/m²K)
Cavity wall (concrete block inner leaf)	1.6
Solid wall (215 mm)	2.3
Pitched roof – cold (tile/slate)	–
Floor (e.g. over unheated garage)	1.9

Source: various

Table 10.7   Limiting *U*-value standards (W/m²K) (ADL1B, Table 1).

Element	Limiting *U*-value
Wall	0.70
Floor	0.70
Roof	0.35
Windows, roof windows, roof lights and doors	3.3

typical 'as existing' elements should be noted. In their unmodified state, thermal elements generally encountered when lofts are converted (such as un-insulated pitched roof slopes and gable walls) will usually fall outside the threshold values and will trigger the guidance on upgrading.

Table 10.6 indicates possible values for un-insulated thermal elements that might be encountered before conversion, although it should be noted that a considerable degree of variation is possible, particularly with cavity walls. In these cases, the elements should be upgraded to the standards set out in Table 10.5 column (b). Where it is not feasible to do this (for example, if it would introduce an unacceptable level of technical risk), a lower standard may be acceptable, but it is emphasised that the values set out in Table 10.7 (*Limiting U-value standards*) represent a last resort.

Note that ADL1B recognises the difficulties inherent in upgrading existing fabric: upgrading should therefore be carried out only if it is technically, functionally and economically feasible to do so. A test of economic feasibility is provided in the guidance, i.e. a simple payback of 15 years (see also *simple payback,*

Glossary). The guidance also suggests a technical argument for lesser provision that would apply where the existing structure was not capable of supporting the weight of additional insulation. ADL1B also indicates that a lesser provision might apply where the thickness of the insulation would reduce useable floor area by more than 5%.

In cases where retained thermal elements are marginally within the threshold values set out in Table 10.5(a) (solid walls 0.69 W/m²K, pitched roof 0.34 W/m²K etc), there would be no requirement to upgrade the fabric, although it would be desirable to do so nonetheless.

Whilst the guidance is based primarily on an elemental approach to compliance, paragraph 18 of the Approved Document indicates that other approaches may be used instead for added design flexibility. These include the use of area-weighted U-values and, where extra latitude is required, the use of the CO_2-based SAP 2005 methodology. In all cases, the limiting U-values set out in Table 10.7 apply.

Standards for renovation of thermal elements

These standards apply when certain types of work are carried out in an *existing* loft conversion or room-in-roof configuration. They are triggered when more than 25% of the surface area of a thermal element is being renovated. The values set out in Table 10.5 column (b) would apply in these cases where feasible.

ADDITIONAL OBLIGATIONS UNDER ADL1B

Providing information

Requirement L1(c) of the Building Regulations (S.I. 2006/652) is that reasonable provision shall be made for the conservation of fuel and power by 'providing to the owner sufficient information about the building, the fixed building services and their maintenance requirements so that the building can be operated in such a manner as to use no more fuel and power than is reasonable in the circumstances.'

The Approved Document indicates that a way of complying would be to provide a suitable set of operating and maintenance instructions aimed at achieving economy in the use of fuel. The instructions should use terms householders can understand and should be in a durable form that can be kept and referred to over the service life of relevant systems. This might include details on adjusting the timing and temperature control settings of a heating system.

Technical risk

Paragraph 7 of ADL1B states that 'The inclusion of any particular energy-efficiency measure should not involve excessive technical risk.' ADL1B refers readers to BR 262 *Thermal insulation: avoiding risks* (BRE 2002). Where the provision of insulating material would introduce an unacceptable level of technical risk, it would be reasonable to limit any upgrading of the thermal element in question to the standard set out in Table 10.7 (*Limiting U-value standards*). It should be noted that this table is provided to act as an end-stop rather than a target: the intention is that any upgrades be limited by technical risk (rather than by Table 10.7) in which case better values might prevail. Note that the minimum acceptable standards in Table 10.7 are intended to reduce the risk of condensation occurring in localised parts of the envelope.

Continuity of insulation and airtightness (new/replacement elements)

There is increased emphasis on minimising thermal bridging and air leakage at gaps, joints and edges. Paragraph 53 of ADL1B directs users to third-party guidance on the subject. Compliance could be demonstrated by a signed report from a suitably qualified person.

U-value calculations

ADL1B indicates that *U*-value calculations shall be carried out using the methods and conventions set out in BR 443 *Conventions for U-value calculations* (BRE 2006).

PRACTICAL MEASURES FOR ENERGY CONSERVATION

The following notes and drawings are intended to provide an outline of some of the practical measures that can be taken to limit heat loss through windows, walls and roofs in loft conversions. Existing and new elements are considered.

In order to keep the structural depth of insulating materials to a minimum, it is practice generally to use rigid foil-faced polyisocyanurate (PIR), polyurethane (PUR) or phenolic foam insulation boards in the majority of applications. These materials offer a high level of thermal resistance relative to their thickness. Note that even relatively modest low-emissivity cavities in insulated constructions (e.g. Fig. 10.2b) make a significant contribution in limiting heat loss from the building envelope.

A continuous vapour control layer should be provided across and between elements. This may take the form of polythene sheeting; equally, it may be integral with proprietary wallboards or rigid insulation materials. Manufacturers' guidance should be followed in all cases.

It is emphasised that this section is intended to provide general guidance only. In individual cases, it will be necessary to carry out a thorough assessment of the existing building fabric; declared performance standards and installation guidance provided by manufacturers should also be taken into account. Note also that the major insulation manufacturers are able to provide technical specifications for specific constructions.

Insulation risks

There are technical risks associated with the provision of insulation, particularly where the insulating material is applied to the inside of an existing building. This is primarily because high levels of insulation make elements outside the insulated envelope much colder and thus more susceptible to the damaging effects of condensation. Effective vapour control (i.e. the provision of contiguous vapour control layers and appropriate ventilation) may limit inside-to-outside vapour movements but, in the case of alterations to existing buildings, such as loft conversions, it not usually possible to eliminate unwanted moisture movements entirely.

Surface condensation

A building with insulation fixed to internal surfaces warms up relatively rapidly but because insulating materials (unlike solid masonry) have a low thermal mass, cooling is correspondingly swift when heating is turned off. This increases the risk of surface condensation occurring. Adequate ventilation will serve to limit the formation of surface condensation. Guidance is set out in Approved Document F *Means of ventilation*.

Interstitial condensation – all elements

Vapour-rich air that escapes from the insulated building envelope will condense in cold structural voids, particularly ones that are not adequately ventilated. Condensation of this sort is potentially destructive over extended periods of time, particularly where it occurs on timber structural members or within insulating materials. Ensuring continuity of the vapour control layer, where possible, will minimise the amount of vapour that escapes from the building; effective ventilation of cold roof structures (see also Chapter 9) will limit the accumulation and condensation of any water vapour that does escape.

Spalling risk – masonry walls

Walls insulated from the inside are no longer warmed by the building and are therefore more likely to be subject to a freeze/thaw regime. Bricks and mortar may spall if moisture is present.

Windows

U-values of windows purchased off-the-shelf may be indicated by the manufacturer. Values for roof windows in ADL1B are based on the U-value having been assessed with the roof window in a vertical position: BR 443 provides guidance on adjustment. New windows or roof windows in the extended area should have a value of 1.8 W/m²K or better (see Table 10.1).

Where data is not available, in the case of some bespoke windows for example, reference may be made to the indicative U-values for windows table in SAP 2005. The configurations in Table 10.8 are based on the SAP table and apply to whole-window constructions (i.e. frame as well as glazing). They are provided for general guidance only.

Table 10.8   Whole-window constructions achieving a *U*-value of 1.8 W/m²K (vertical).

Frame	Inter-pane gap	Void fill	Glazing
Wood/PVC-U	6 mm	Argon	Triple, low-E ($\varepsilon_n = 0.15$) hard coat
Wood/PVC-U	12 mm	Argon	Double, low-E ($\varepsilon_n = 0.05$) soft coat
Wood/PVC-U	16 mm	Air	Double, low-E ($\varepsilon_n = 0.05$) soft coat
Metal*	6 mm	–	–
Metal*	12 mm	Argon	Triple, low-E ($\varepsilon_n = 0.05$) soft coat
Metal*	16 mm	Air	Triple, low-E ($\varepsilon_n = 0.05$) soft coat

* 4 mm thermal break assumed

It should be noted that loft conversions are prone to overheating in hot summer weather. Roof windows positioned at a relatively high point on the roof slope will provide some assistance in controlling the build-up of heat. By contrast, low level windows are less effective in aiding the dispersion of hot air that accumulates at apex or ceiling height (see also *Purge ventilation* below).

Existing solid brick masonry walls (*U*-value 0.35 W/m²K)

As noted earlier in this chapter, solid masonry walls are poor insulators. An unmodified 9″ (225 mm) wall generally has a *U*-value of about 2.3 W/m²K (Fig. 10.2a), or worse. Where a wall of this sort is to be incorporated within the thermal envelope of the conversion, it must generally be upgraded to conform to the improved value (0.35 W/m²K) set out in ADL1B.

Generally, the only practical way of doing this is to provide insulation to the inside of the wall. Fig. 10.2b illustrates one method. It is also possible to fix insulation outside the building envelope, but this is usually not practical unless the whole height of a wall is similarly treated.

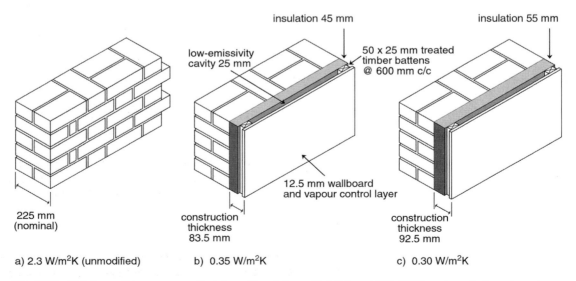

a) 2.3 W/m²K (unmodified)   b) 0.35 W/m²K   c) 0.30 W/m²K

Fig. 10.2   Existing solid masonry walls: internal insulation with foil-faced PIR, PUR or phenolic foam boards.

New solid brick masonry walls (*U*-value 0.30 W/m²K)

New solid masonry walls may be constructed when an existing gable wall is extended to form a flank gable, or when a hip-to-gable conversion is undertaken. In these cases, the thickness of the new wall is limited to that of the existing wall structure for practical reasons. The standard for new walls is 0.30 W/m²K, slightly higher than that for existing walls.

As with existing walls, the most practical method is to provide insulating material inside the building. Where both old and new elements are incorporated in a single wall structure (for example, in a flank gable construction) there would be little to be gained by adopting two different standards of insulation on adjoining sections. Note that the thickness of construction required to achieve a *U*-value of 0.30 W/m²K is only marginally greater than that needed for 0.35 W/m²K (Fig. 10.2c).

New solid blockwork walls (*U*-value 0.30 W/m²K)

A new gable may be built off the existing walls using solid AAC (autoclaved aerated concrete) blockwork when a hip-to-gable conversion is undertaken. This approach may be adopted for buildings with either existing cavity or solid masonry walls. Manufacturers produce blocks in thicknesses that co-ordinate with both 9" (225 mm) solid walls and 'traditional' 10½" (265 mm) cavity configurations.

Either lightweight or ultra-lightweight aerated concrete blockwork may be used. Typically, lightweight blocks have a thermal conductivity of approximately 0.18 W/mK; ultra-lightweight blocks are more effective insulators with a conductivity value of about 0.11 W/mK.

The advantages of constructing walls in solid blockwork are speed and cost. The main disadvantage is that aerated concrete blocks alone will not generally achieve the new-wall *U*-value (0.30 W/m²K) set out in ADL1B without the addition of supplementary insulation.

However, it is still possible to achieve results that are significantly better than the overall limiting value for walls (0.70 W/m²K). Fig. 10.3 illustrates 215 and 265 mm blockwork walls in both modified and unmodified configurations. Note that the compressive strength of lightweight and ultra-lightweight blockwork should be considered where beam bearings are to be introduced.

Existing cavity masonry walls (*U*-value 0.55 W/m²K)

In a dwelling with cavity walls, the cavity generally continues into the gable structure although it should be noted that this is not universally the case: even in recently constructed dwellings, the gable above ceiling height at existing top floor level is sometimes constructed in solid masonry.

As noted above, the overall thickness of the 'traditional' cavity wall is generally 10½" (265 mm). Common early versions (pre-WWII) generally comprise an outer brick leaf, a cavity (usually a nominal 2"/50 mm) and an inner leaf in brick, clinker

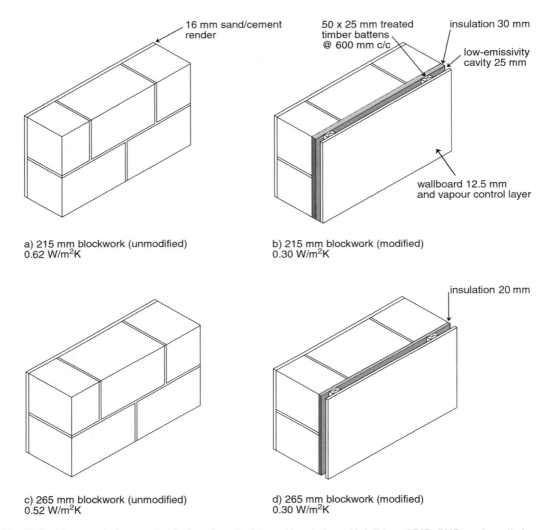

a) 215 mm blockwork (unmodified)
0.62 W/m²K

b) 215 mm blockwork (modified)
0.30 W/m²K

c) 265 mm blockwork (unmodified)
0.52 W/m²K

d) 265 mm blockwork (modified)
0.30 W/m²K

Fig. 10.3 New aerated concrete blockwork walls: internal insulation with foil-faced PIR, PUR or phenolic foam boards (block conductivity: 0.11 W/mK).

or concrete block. The thermal performance of such walls varies. With an inner leaf in medium-density concrete blockwork, an overall U-value of about 1.6 W/m²K would be typical. Better overall values are achieved when the inner leaf is constructed in clinker block (1.4 W/m²K).

If the cavity has been filled with insulating material (blown mineral wool or urea formaldehyde foam for instance), the wall will achieve a considerably better U-value. In the case of a brick inner leaf, cavity fill would provide an overall value of approximately 0.58 W/m²K.

In the post-war period, aerated concrete blocks started to take the place of dense concrete, clinker and brick for inner-leaf construction. For 50 mm cavities, walls with a 100 mm AAC blockwork inner leaf achieve a U-value of about $0.95\,\text{W/m}^2\text{K}$ if they have not been filled and $0.47\,\text{W/m}^2\text{K}$ if they have.

Retro-fit cavity wall insulation was introduced in the late 1960s and cavity insulation for new build in England and Wales became routine from the mid-1980s. However, even where cavity insulation has been provided, it should not be assumed that it extends to full gable height: it may be necessary to 'top up' cavity insulation to bring it up to the apex. If this approach is adopted, the 'top up' insulant should match the existing material. For practical reasons, this approach could only reasonably be applied with materials such as blown mineral wool, EPS beads or urea formaldehyde (UF) foam.

An alternative and sometimes less troublesome approach for existing cavity walls is to apply insulating material to the inner face of the wall. Fig. 10.4 illustrates some common cavity configurations.

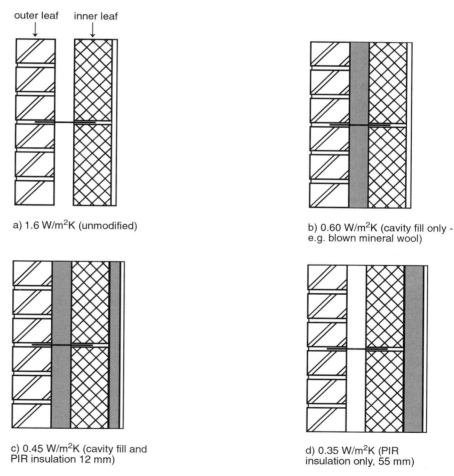

a) $1.6\,\text{W/m}^2\text{K}$ (unmodified)

b) $0.60\,\text{W/m}^2\text{K}$ (cavity fill only - e.g. blown mineral wool)

c) $0.45\,\text{W/m}^2\text{K}$ (cavity fill and PIR insulation 12 mm)

d) $0.35\,\text{W/m}^2\text{K}$ (PIR insulation only, 55 mm)

Fig. 10.4 Existing cavity walls: nominal 100 mm brick/50 mm cavity/100 mm medium-density concrete blockwork.

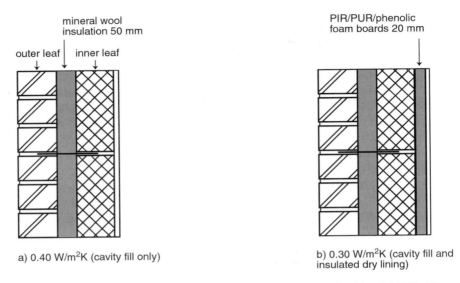

a) 0.40 W/m²K (cavity fill only)

b) 0.30 W/m²K (cavity fill and insulated dry lining)

mineral wool insulation 50 mm

PIR/PUR/phenolic foam boards 20 mm

outer leaf | inner leaf

Fig. 10.5   New cavity walls: aerated concrete blockwork (block conductivity: 0.11 W/mK).

New cavity masonry walls (*U*-value 0.30 W/m²K)

The structural thickness of the existing wall will determine the thickness of the new section and, in most cases, this will be 265 mm. It is possible to achieve a *U*-value of approximately 0.40 W/m²K with this thickness of structure by constructing the inner leaf from ultra lightweight blockwork and providing cavity fill, such as mineral wool, as the wall is raised. However, in order to achieve the new-wall *U*-value of 0.30 W/m²K, some supplementary insulation to the inner face would normally be needed (Fig. 10.5).

New tile hung stud walls (*U*-value 0.30 W/m²K)

New stud walls must be insulated to the same standard as new masonry walls. Insulation, generally PIR, PUR or phenolic foam boards, is friction-fitted between the studs and an additional internal layer of insulation is also provided. Note that, in Fig. 10.6a, the construction illustrated provides a lower standard of insulation than that shown in Fig. 10.6b even though the overall thickness of insulating material is greater.

Existing or new pitched roof (*U*-value 0.20 W/m²K)

The standard for insulating both new and existing pitched roof slopes is the same under guidance set out in ADL1B. Fig. 10.7 illustrates one method of achieving a *U*-value of 0.20 W/m²K with a typical tiled roof structure with 47 × 100 mm rafters at 400 mm c/c. Note that the depth of construction projecting beneath the soffit of the rafters is in excess of 100 mm in this configuration. The additional loading created by insulation, internal battens and wallboard must also be considered.

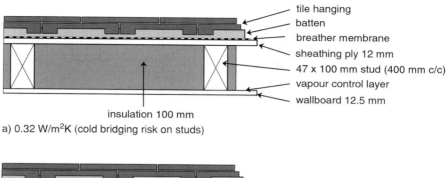

insulation 100 mm

a) 0.32 W/m²K (cold bridging risk on studs)

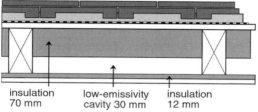

| insulation 70 mm | low-emissivity cavity 30 mm | insulation 12 mm |

b) 0.30 W/m²K (cold bridging minimised)

Fig. 10.6 New tile-hung stud wall: insulation with foil-faced PIR, PUR or phenolic foam boards.

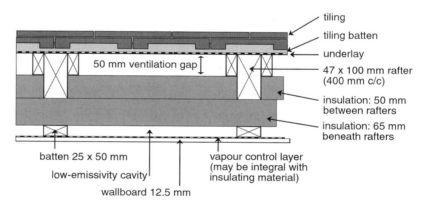

Fig. 10.7 Pitched roof insulation to achieve 0.20 W/m²K with foil-faced PIR, PUR or phenolic foam boards.

New flat warm roof (*U*-value 0.20 W/m²K)

Fig. 10.8 illustrates one method of achieving a *U*-value of 0.20 W/m²K with a warm deck flat roof using PIR, PUR or phenolic foam boards. Boards with surfaces designed to withstand the application of hot-bedded roofing must be used where a built-up finish is to be applied. Note the considerable depth of structure created in this configuration: in a typical loft conversion dormer flat roof spanning 3 m, the depth of construction *above* the joists (including firring) would exceed 220 mm at the highest point.

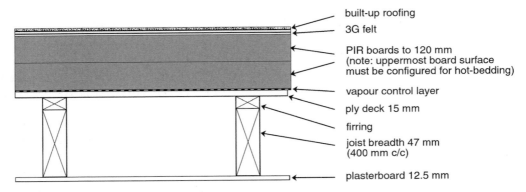

Fig. 10.8   Flat roof (warm deck) to achieve 0.20 W/m²K with PIR-type insulation.

New flat cold deck (*U*-value 0.20 W/m²K)

Fig. 10.9 illustrates two possible cold deck configurations achieving 0.20 W/m²K. Note that, in all cold roof structures, it is necessary to maintain a 50 mm ventilation gap above the insulation material – the depth of firring should therefore be discounted unless it is greater than 50 mm at its thinnest end.

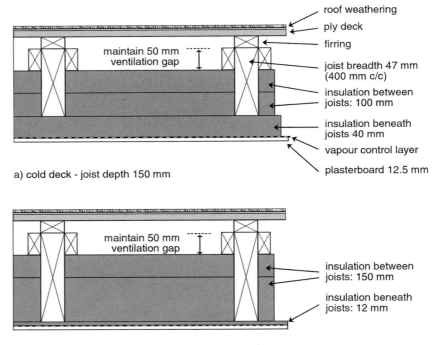

Fig. 10.9   Flat roof (cold deck) to achieve 0.20 W/m²K with PIR, PUR or phenolic foam boards.

VENTILATION FOR OCCUPANTS

Guidance on providing means of ventilation for people in buildings is set out in Approved Document F. Note that guidance on condensation in roofs, formerly contained in Approved Document F2, has been moved to Approved Document C *Site preparation and resistance to contaminants and moisture* (2004). In addition, it is emphasised that air supply for combustion appliances should be considered in relation to Part J of the Building Regulations and guidance in Approved Document J.

The emphasis on airtightness in Approved Document L *Conservation of fuel and power* has increased the importance of accounting for air exchanges through the building envelope. The 2006 edition of Approved Document F, therefore, adopts a more comprehensive approach than earlier versions, and purpose-provided ventilation is considered in detail. Note that the 2006 guidance introduces some new terminology:

- *Purge (rapid) ventilation.* Manually-controlled high-rate ventilation. Typically provided by openable windows but may be provided by a fan instead.
- *Whole building (background) ventilation.* Nominally continuous low-rate ventilation of rooms. Typically, this is provided by manually controlled trickle vents in window frames. Wall ventilators may also fulfil this function. In all cases, it is recommended that these be positioned 1.7 m above floor level to limit the perception of cold draughts. Note that 'equivalent area' instead of 'free area' is now used for sizing background ventilators, including trickle ventilators. The method for measuring equivalent area (which measures the airflow performance of the ventilator) is set out in BS EN 13141-1:2004. The old 'free area' (geometric measure) of a trickle ventilator is approximately 25% larger than the new 'equivalent area'.

  Section 3 of the Approved Document indicates that there should be an undercut of minimum area 7600 mm^2 to internal doors between a wet room and the existing building (equivalent to 10 mm depth for a standard 760 mm width door).
- *Extract ventilation.* In dwellings, extract ventilation is provided in rooms where higher levels of water vapour and pollutants, including odour, are encountered. Extract ventilation may be mechanical (e.g. fan) or natural (e.g. passive stack ventilation).

Ventilation for habitable rooms

Whole-building (background) ventilation

Habitable rooms such as bedrooms in a loft conversion should each be provided with whole-building (background) ventilation with an equivalent area of 5000 mm^2. This is generally provided by trickle ventilators incorporated into window assemblies.

Purge (rapid) ventilation

Purge ventilation is required in each habitable room and is usually provided by openable windows (including roof windows). For a hinged or pivot window that opens 30 degrees or more, or for sliding sash windows, the area of the opening part should be at least 5% of the floor area of the room. For a hinged or pivot window that opens less than 30 degrees, the area of the opening part should be at least 10% of the floor area of the room.

In habitable rooms with more than one openable window, the areas of all opening parts may be summed to achieve the required proportion of floor area, with the proportion of floor area (i.e. 10% or 5%) determined by the opening angle of the largest window in the room. Note that the minimum size requirements for escape windows must be observed in all cases (see Chapter 4).

Extract ventilation

This is generally not required in habitable rooms such as bedrooms.

Ventilation for wet rooms (includes bathrooms and WCs)

Section 3.12 of Approved Document F sets out four strategies for providing whole building and extract ventilation when a wet room (such as a bathroom or WC) is added to an existing building. The following approach may be considered.

Whole-building (background) ventilation

For bathrooms and WCs, background ventilation of at least 2500 mm^2 equivalent area is required.

Purge (rapid) ventilation

An openable window may be installed in an external wall; no minimum size is specified. For wet rooms with no external wall, see below.

Extract ventilation

Minimum intermittent extract rate of 15 litres/second (bathroom) and 6 litres/second (WC). For a room with no openable window and no natural light, an extract fan should have an overrun of 15 minutes and would generally be triggered by the operation of the light switch. For dedicated WCs only, purge ventilation provisions may be adopted instead of extract ventilation.

Appendix 1 Specification

Drawings and plans are a basic requirement for any building work. But the form and volume of technical documentation produced prior to construction varies from project to project and depends to a great extent on how, and by whom, the work is to be carried out. An architect, for example, might provide a comprehensive specification as well as a detailed schedule of works: these, in tandem with drawings, would provide the basis for a contractor's quotation.

In the case of specialist loft conversion firms, however, contractor and designer are one and the same in many instances. The level of technical detail contained in a specialist's specification may therefore be limited to indicating compliance with building regulations and planning requirements: specification or construction notes may simply take the form of extended annotations to drawings.

Specification of certain proprietary items (VELUX windows with model numbers, Celotex insulation with product code) will help to indicate compliance. Where the specification relates to a system rather than a single specific product (a built-up felt roof, for example), indicating the materials to be used, the mode of application and a relevant standard would be appropriate.

The following specification is based on a typical hip-to-gable loft conversion carried out by a specialist conversion company under permitted development during 2005. The conversion adds a new storey to an existing two-storey dwelling. The storey area is less than 50 m^2. Two habitable rooms and a bathroom with WC are provided. This specification is provided for the purposes of illustration only. References to conservation of fuel and power (insulation/energy-efficient lighting) and ventilation for occupants are based on pre-2006 guidance.

Beams

- Install universal beams and universal columns in accordance with structural engineer's design. Beam ends to bear on padstones or mild steel bearing plates. Plates or padstones to be built into brickwork and bedded in 10 mm thick mortar. Reinstate brickwork around beam ends and pack voids well with mortar.
- Floor beams to provide 30-minute fire resistance by coating with intumescent paint or by encasing in 2 no. layers of 12.5 mm plasterboard with staggered joints taped and filled. All new beams to stand 25 mm (minimum) clear of existing ceiling.

New floor

- Fix new 47 × 195 mm strength class C16 structural floor joists at 400 mm c/c. Joist ends to be fixed directly into beam webbing with solid web blocking, or fixed to beam bearers or trimmers with Simpson Strong-Tie JHA450 hangers fully nailed. Solid strutting to be provided between joists spanning in excess of 2500 mm. Joists to be 40 mm clear of chimney breasts or 200 mm clear of flue. Double joists to be provided under non load-bearing partitions. Multiple timbers bolted at 600 mm c/c with M12 grade 4.6 bolts, washers and toothed plate connectors.

- Existing ceiling joists to be strapped to new floor joists with 20 mm × 1 mm galvanised steel band.
- 100 mm mineral wool between existing ceiling joists supported on chicken wire where existing ceiling is lath and plaster or 9.5 mm plasterboard. Mineral wool to be run up to but not into eaves. Where existing ceiling is 12.5 mm plasterboard, chicken wire omitted.
- Lay 22 mm T&G flooring-grade chipboard throughout, including eaves storage space over existing ceiling joists. Floor to bathroom to be laid with moisture-resistant chipboard as specified in BS 7331:1990 or BS EN 312 Part 5:1997 to conform to Approved Document C (2004).

Walls – dormer

- 47×100 mm vertical studs (C16) fixed to sole and head plates and header beams. External sheathing to be 11 mm OSB3 cloaked with breather membrane.
- 38×25 mm treated softwood tiling battens minimum length 1200 mm to be fixed through membrane and OSB to studwork at 114 mm gauge (maximum) with plain tile hanging (265×165 mm). Fix soakers and flashings as required to all roof and window junctions in code 4 lead sheet.
- Where external stud wall has an area greater than 1 m^2 and is within 1000 mm of a relevant boundary, fix 9 mm Promat Supalux boarding between sheathing material and underlay/tile battens to the outside, and finish internally with 12.5 mm plasterboard with 3 mm gypsum plaster finish.
- Friction fit 2 no. layers of Celotex GA3050Z rigid insulation to all stud voids with polythene vapour barrier to warm side of insulation. Finish with 12.5 mm plasterboard with plaster skim coat.

Wall – hip to gable (solid masonry)

- Existing hip and jack rafters to be removed and replaced with new 47×100 mm common rafters at 400 mm c/c.
- New 225 mm gable wall in stock bricks to match existing with 50 mm Gyproc ThermaLine Super thermal laminate internally on dabs with skim plaster finish.
- Re-cover new pitched portion of roof with tiles recovered from hip on underlay on battens.
- Restraint straps (30×5 mm) to be fixed to gable end and to three rafters/floor joists at 2 m (maximum) spacing.

Walls – purlin and perimeter

- Purlin and perimeter walls to be 47×100 mm vertical studs (C16) at 400 mm c/c or centres suitable to rafters with 47×100 mm head and base plates. Stud voids to be filled with 2 no. layers of Celotex GA3050Z rigid insulation. 12.5 mm plasterboard with plaster skim coat finish.

Walls – internal partitions

- Internal partitions to be 47×100 mm vertical studs at 400 mm c/c fixed to head and base plates with staggered noggings. Both sides faced with 12.5 mm Gyproc SoundBloc (minimum mass 10 kg/m^2) and plaster skim coat finish. Voids to be filled with snug-fitting 100 mm mineral wool to conform to guidance in Approved Document E (2003).

Walls – stair enclosure

- 47×100 mm vertical studs at 400 mm c/c fixed to head and base plates with staggered noggings. Both sides faced with 12.5 mm plasterboard with plaster skim coat finish to provide 30-minute fire resistance.

Roof windows

- Roof window openings to be fully trimmed with 2 no. 47×100 mm C16 rafters. Trimmers also provided above and below opening. All windows to be double glazed with 4/16/4 low-E glass and installed in accordance with manufacturer's instructions.

Escape window

- Habitable rooms to have access to emergency egress window with minimum unobstructed openable area of at least 0.33 m^2 and with minimum dimensions of at least 450×750 mm e.g. VELUX GPL M08. Escape window to be 1700 mm (maximum) from eaves measured along roof slope. Bottom of openable areas to be between 600 mm and 1100 mm above finished floor level.

Windows and ventilation

- All windows to be double-glazed using 4/16/4 low-E glass to *U*-value of 0.20 W/m^{2}K (Approved Document L, 2002). Guarding to be provided where sill height is less than 800 mm above finished floor level unless emergency egress window in roof slope.
- All habitable rooms to have openable windows with rapid ventilation equivalent to 5% of floor area with part minimum 1750 mm above floor level. Background ventilation of 8000 mm^2 (minimum).

- Bathroom to be fitted with mechanical extractor to achieve 15 litres/second extraction connected to light switch and set to overrun for 15 minutes if window not provided. Bathroom to have background ventilation of 4000 mm^2.

Existing roof slope – insulation

- 12.5 mm plasterboard on polythene vapour barrier on 35 mm Celotex GA3035Z rigid insulation fixed to rafters with Celotex GA3050Z fitted between rafters. 50 mm air gap to cold side of insulation. *U*-value 0.3 W/m^{2}K.

Existing roof – ventilation

- Existing roof structure to be provided with proprietary ventilators or vents to existing eaves to achieve equivalent continuous ventilation of 25 mm (eaves). Tile or ridge vents to give equivalent of 5 mm continuous ventilation (ridge). Provide insect mesh to all vents.

Flat roof – warm deck

- 12 mm mineral chippings bedded in bitumen on 3 no. layers of polyester base roofing felt (BS 747) hot bedded and laid to BS 8217. Base layer to be single-ply G3 felt with 25 mm diameter holes equally spaced and partially bonded to Celotex TD3080Z warm deck insulation boards bedded on mastic beads at joints according to manufacturer's guidance to achieve vapour check. *U*-value 0.25 W/m^{2}K.
- Roof deck to be 18 mm exterior plywood on tapered firrings to achieve 1:40 fall on 47 × 170 mm flat roof joists (C16) at 400 mm c/c. Flat roof joists to be fixed into roof beam webbing. Roof joists at dormer face to be fixed with holding-down straps to window header (2 no. 47 × 170 mm). Ceiling to be finished internally with 9.5 mm plasterboard and skim coat plaster.

Flat roof – cold deck

- 12 mm mineral chippings bedded in bitumen on 3 no. layers of polyester base roofing felt (BS 747) hot bedded and laid to BS 8217. Base layer to be single-ply G3 felt with 25 mm diameter holes equally spaced and partially bonded to roof deck.
- Roof deck to be 18 mm exterior plywood on tapered firrings to achieve 1:40 fall on 47 × 170 mm flat roof joists (C16) at centres to match existing rafters. Celotex GA3090Z rigid insulation to be fitted in roof joist voids with minimum 50 mm air gap to cold side. Flat roof joists to be fixed to top of ridge beam to provide continuity of ventilation. Joists at dormer face to be fixed with holding-down straps to window header (2 no. 47 × 170 mm). Fascia or soffit vents to provide equivalent of 25 mm continuous ventilation. Ceiling to be finished internally with 9.5 mm plasterboard and skim coat plaster.

Flat roof – general

- Fix softwood/PVC-U soffit and fascia boards with 112 mm gutter set to 1:350 fall to 68 mm downpipe discharging to existing rainwater drainage system.

Staircase

- Estimated floor-to-floor of 2756 mm gives 13 equal risers of 212 mm and equal going of 245 mm at 42-degree (maximum) pitch. Width 750 mm. 12.5 mm plasterboard to stair soffit.
- Headroom minimum 1800–2000 mm depending on position to conform to Approved Document K.
- Handrail set at 900 mm minimum above pitch line. No part of balustrade or staircase to allow passage of a 100 mm sphere, to conform to requirements of AD K.

Electrical

- Provide one fixed energy-efficient light fitting to achieve a luminous efficiency greater than 40 lumens per circuit watt.
- Electrical installation to be designed, installed, inspected and tested in accordance with the requirements of BS 7671 (IEE Wiring Regulations) and Approved Document P *Electrical safety*.

Drainage

- Bath, basin and shower waste pipes to be 40 mm diameter. Runs up to 4 m to be 50 mm diameter all connected separately with water seal traps to existing SVP or new 100 mm common branch pipe (1:40 fall). Cleaning access to be provided at change of pipe direction. Opposed connections to SVP to be offset at least 200 mm.
- Vent pipe within 3 m of any openable window to be extended 900 mm above window and provided with cage or perforated cover to conform to guidance in Approved Document H.

Fire

- All new doors to provide fire resistance to FD 20* standard with self-closing devices and intumescent strips. Door stops 25 × 35 mm glued and screwed. Existing internal doors to be fitted with self-closing devices.
- Where built over habitable rooms, circulation spaces with existing 18 mm plain edge floor boards to be overlaid with 4 mm hardboard.

* Note that while the FD 20 standard is still referenced in Approved Document B Fire safety, it is no longer widely available, and FD 30 doors and linings are used instead.

- Door glazing and fanlights within protected stair enclosure to be fitted with fire-resisting glazing.
- Provide mains-operated linked smoke alarms fitted with battery reserve supply to all floors to conform to guidance in Approved Document B.

Water tanks

- Provide replacement water storage tank with valves, tank lid and pipework as necessary.
- Remove and reposition central heating feed and expansion tank.

Appendix 2 The Building Regulations: appeals and determinations

The appeal and determination mechanisms described in Chapter 2 may be employed when a dispute arises between a building control body and a person carrying out building work. Recourse to either of these procedures is generally considered to be a last resort and outcomes seldom favour the applicant, although there are some notable exceptions to this. The number of appeals and determinations is relatively small, with fewer than 100 cases considered in the six years to the end of 2004 in England and Wales.

However, more than 20% of cases in this period were concerned with aspects of loft conversion in single-family dwellinghouses. This is a high proportion when loft conversions are considered as part of construction activity as a whole and to an extent this reflects the difficulties encountered when adapting existing structures for new purposes.

For reasons described elsewhere in this book, providing appropriate means of escape and safe stair access is not always straightforward in the case of loft conversions. It is therefore not surprising that most of the appeals and determinations concerning loft conversions deal with these requirements. Three-quarters of the cases concern Requirement B1 *Means of warning and escape*. This emphasises the sometimes intractable problems of internal layout, but arguments advanced by applicants and appellants also indicate a widespread misunderstanding of Requirement B1 and particularly the guidance that supports it.

Determination and appeal decisions should not be regarded as setting a precedent. The Secretary of State is required to consider all cases on their individual merits and some factors that are specific to previous cases are not relevant to subsequent ones.

However, the arguments advanced in both the determinations and the appeals procedures provide a valuable insight into the way in which Building Regulations requirements and the supporting guidance are interpreted. Decisions and determinations by the Secretary of State thus serve to reinforce and sometimes clarify the application of existing guidance.

As explained in Chapter 2, the determination and appeals procedures have distinct roles. In a determination, a decision is made on whether or not a proposal complies with specific Building Regulations requirements. In an appeal, the appellant accepts that a proposal does not comply and seeks a relaxation or dispensation of a requirement.

The following pages contain extracts from determination and appeal letters published by the Secretary of State all of which concern loft conversions. Where appropriate, the letters are reproduced in full.

(1) Open-plan layout at ground floor
(2) Sprinkler systems and open-plan layouts
(3) Potentially habitable rooms and enclosure
(4) Doors of unknown fire resistance
(5) Open-plan layout – escape windows
(6) Definition of a stair
(7) Stair design – pitch and going
(8) Stair design – selective application of regulations
(9) Means of escape – floors more than 7.5 m above ground level
(10) Means of escape – ladder brackets
(11) Means of escape – structural constraints
(12) Sleeping gallery

(1) OPEN-PLAN LAYOUT AT GROUND FLOOR

An open-plan layout at ground floor level is seldom compatible with escape provisions in dwellings where the conversion creates a three-storey dwelling or higher. The arguments advanced by the Secretary of State in the following determination of compliance with Requirement B1 (reference 45/1/214 dated 15 September 2004) are representative of those put forward in a number of other similar cases.

The proposed work

The proposed building work comprises alterations to, and the extension of, an existing two-storey house including the conversion of its roof space to a bedroom, i.e. a loft conversion, thereby creating a second floor (third storey). The existing plan dimensions of the house are approximately 4.5 m (frontage) × 7 m.

The extension will provide a new kitchen (approximately 2.5 m × 4.25 m) at ground floor level and a new bedroom (approximately 2.5 m × 2.25 m) at first floor level provided with a window suitable for escape. Existing partition walls at ground level floor are also to be removed to create a single 'open plan' living room running from the front door to the new kitchen and opening onto a small courtyard at the rear. At first floor level a remodelled layout will also provide a new en suite bathroom for the master bedroom and a separate bathroom.

The new bedroom on the second floor is approximately 4 m × 3.5 m. It is to be created by breaking open the rear pitch of the roof and the installation of a wide dormer window of 2.7 m in width. The room will also be served by three roof windows in the front elevation, one of which is annotated as suitable for escape.

A new stair serving each floor is to be provided starting from the rear of the living room adjacent to the kitchen door. All doors leading on to the stairway from the first and second floors and the kitchen door are to be fire doors.

The plans also indicate that two mains operated interlinked smoke alarms are proposed in the living room area and one on each of the first and second floor landings; and that a heat alarm is proposed for the kitchen.

These proposals were the subject of a full plans application which was rejected by the Borough Council. Two subsequent re-submissions of the full plans were also considered but these too were rejected on the grounds of failure, amongst other things, to comply with Requirement B1. The Council took the view that your proposals were not a suitable substitute for providing a protected escape route at ground floor level for the stair. However, you contend that your proposals do comply with Requirement B1 and it is in respect of that question that you have applied to the Secretary of State for a determination.

The applicant's case

You indicate that the guidance in Approved Document B *Fire safety* only provides examples of how compliance with Requirement B1 can be achieved and that these are not exclusive.

You refer to the following proposed provisions which you consider will provide prompt, safe and adequate means of escape in the event of fire:

(i)   A mains-operated system of interlinked smoke alarms at each floor level plus a heat alarm in the kitchen to give early warning of fire.
(ii) The stair will be protected at first and second floors with 30-minute fire resisting construction. All doors are to be self-closing fire doors with smoke seals.
(iii) A self-closing fire door will also be provided to the kitchen on the ground floor, thus protecting the stair from a possible fire from the kitchen.
(iv) There are two alternative exits from the house at ground level.
(v)   Escape windows will be provided at first and second floor levels as an alternative means of escape, as recommended in Approved Document B.

You conclude with your view that the building is not particularly high – the second floor level is 5.125 m above ground floor level – and that 'open plan' arrangements, such as that proposed on the ground floor, are a common feature in small terraced houses.

The Borough Council's case

The Borough Council takes the view that your proposals do not comply with Requirement B1 because they do not meet the criteria recommended in paragraph 2.18 of Approved Document B, relating to loft conversions in an existing two-storey house.

The Borough Council considers that the provision of smoke detection at all levels and a rescue window at second floor level are already part of a set of minimum provisions for loft conversions. The Council contends that your proposals are not a suitable substitute for providing a protected route at ground floor level whereby the stair at first and second floor levels – as the principle means of escape – could be prejudiced by smoke and fire at ground floor level, as the stair will discharge directly into the living accommodation.

The Secretary of State's consideration

The Secretary of State notes that in this case a new room is proposed within the roof space of an existing two-storey house. This has, in effect, created an additional storey which is more than 4.5 m above ground level when measured from the lowest ground level. At this height it is not considered to be safe for people to make their own escape from windows. Therefore, it would normally be necessary to provide a protected escape route down through the house formed with fire-resisting construction and fire-resisting self-closing doors.

Approved Document B provides an alternative strategy for means of escape in these situations specifically for loft conversions in two-storey houses. Using this approach, existing doors need not always be replaced but are just made self-closing, providing a degree of protection for the escape route from the loft room to the final exit. Because the protection to the escape route may not be as effective as that provided in a new house, however, fire-resisting construction is also necessary to separate the new accommodation from the rest of the house. This is intended to allow the occupants of the new floor to wait, in relative safety, for rescue via a ladder through a suitably sized and positioned window.

The Secretary of State takes the view that escape via a stairway protected in this manner would, in the majority of cases, be possible and that waiting for rescue from the window in the loft room should be a last resort. However, in this case, the escape route for the occupants of the loft room is not protected at ground floor level and the likelihood that they might need to be rescued is greatly increased.

You have, however, argued that the provision of fire detection at each floor level, escape windows at first and second floor levels and a fire-resisting enclosure of the stairway at the upper levels of the house provide an adequate level of safety. The Secretary of State considers that these provisions are essentially what would normally be provided if the escape route was protected throughout its length. As such, these provisions could not be considered as mitigating the absence of protection at ground floor level and your proposals do not therefore demonstrate compliance with Requirement B1.

The determination

The Secretary of State has given careful consideration to the particular circumstances of this case and the arguments presented by both parties.

As indicated above, the Secretary of State considers that your proposals as submitted do not make appropriate provision for early warning and means of escape in case of fire from the proposed loft room at second floor level. He has therefore concluded and hereby determines that your proposals do not comply with Requirement B1 *Means of warning and escape* of Schedule 1 to the Building Regulations 2000 (as amended). You should note that the Secretary of State has no further jurisdiction in this case.

(2) SPRINKLER SYSTEMS AND OPEN-PLAN LAYOUTS

Sprinkler systems are increasingly used to enhance fire safety in dwellings. However, the installation of such systems alone is generally not regarded as a substitute for a protected or enclosed escape route where a loft conversion is concerned and a three-storey dwelling (or higher) is created. This is evidenced by a number of determinations in which applicants have proposed the use of sprinkler systems in order to retain an open plan layout. In all cases, it was determined that the proposals did not comply with Requirement B1.

An appeal against refusal to dispense with Requirement B1 (reference 45/3/148 dated 26 July 2001) considered the retention of an open plan layout at ground floor level. The proposal incorporated a high level of both passive and active fire protection measures, including the use of 60-minute fire-resisting construction, new fire-resisting self-closing doors and an integrated fire detection and alarm system in conjunction with a domestic sprinkler system.

Whilst the appeal was dismissed, it should be noted that the basis of the dismissal was procedural rather than technical. The Secretary of State considered that the principle of the proposals had the potential to comply with Requirement B1 and there was thus no prima facie case for the need to relax or dispense with the requirement.

(3) POTENTIALLY HABITABLE ROOMS AND ENCLOSURE

As noted above, problems frequently arise where a stairway discharges directly into an open-plan layout, effectively a habitable room, at ground floor level. However, it is stressed that the relationship between the stairway and habitable rooms may become critical *elsewhere* in a dwelling as well. For example, the need to provide space for a new stair to a loft conversion means it is sometimes necessary to reconfigure rooms and landings on the floor immediately below the new storey.

In determination 45/1/194 (1 June 2001), the Secretary of State considered, amongst other things, the possible use of a large floor area (5 × 3 m) at the head of a first floor stair. The applicant's client contested that this space could only function as a landing. Whilst noting that there was no definitive way of deciding whether or not the area would be regularly used for habitable purposes, the Secretary of State considered that:

'. . . it is a reasonable assumption that the space is likely to be used as more than a landing and therefore, as an open-plan first floor layout, could present a considerable fire risk and threat to the occupants of the building.'

A similar principle applied in determination 45/1/215 (13 August 2004). Because it was not possible to position a new stair to a loft conversion above the existing stairway, it was proposed to locate it within a first floor room previously used as a bedroom with separation from the rest of the first floor provided by a new self-closing 30-minute fire-resisting door. The applicant indicated that this room, with approximate measurements of 4.5 × 3.5 m, would be used as a dressing room and would not be used as a bedroom independently of the proposed second floor accommodation. In addition, the applicant suggested that the case rested on whether there is a maximum size for an enclosure in which the stair may be located and whether the provision of wardrobes or cupboards within a stairway alters the space from an enclosure to a habitable room.

The Secretary of State acknowledged that there will often be some form of fire loading within circulation routes in domestic situations, but that the risk of a fire starting increases when people are engaged in activities other than simply travelling from one room to another. In terms of making a judgement as to whether a space should be regarded as part of a protected stairway, or as a room likely to be regularly used for habitable purposes, the Secretary of State commented that:

'. . . some guidance can be derived from the scale of the building; the number of rooms and the usability of the space; and the number and position of the doors which open off the area. In this case the dressing room is similar in size to the adjacent bedrooms at first floor level and, as a dressing room, would clearly be used for purposes other than circulation.'

(4) DOORS OF UNKNOWN FIRE RESISTANCE

Guidance provided in Approved Document B *Fire safety* indicates that, in certain situations, existing doors may be retained provided that they are made self-closing. However, this guidance requires that any new doors, and any door (or doors) that provides part of the fire separation for a loft conversion, must offer full 30-minute fire resistance.

In an appeal against refusal to dispense with Requirement B1 (reference 45/3/127 dated 23 July 1999), the appellant proposed the retention of a period panel door at first floor level. This door provided access to the base of a new stair for a loft conversion on the floor above. Because the new stair was to be open to the new loft room, the period panel door at first floor level would necessarily form an integral part of the fire separation of the conversion and would require 30-minute fire resistance.

The appellant had been careful to conserve and restore the period fittings of the Victorian house and believed that the provision of a new fire door would be inappropriate under the circumstances.

The appeal was dismissed on the grounds that Requirement B1 is a life safety matter and it is not normally considered appropriate to either relax or dispense with it. However, the Secretary of State expressed the view that there were possible methods of demonstrating compliance with B1. These included an independent assessment of the fire-resistance of the door by a recognised body and, alternatively,

> '. . . it is possible to upgrade the fire resistance of doors of architectural merit by the use of an intumescent coating to an appropriate thickness. However, again the acceptability of this solution in terms of compliance with Requirement B1 would be a matter between you and the District Council.'

Note that a publication by TRADA – *Fire resisting doorsets by upgrading* – deals with this subject and is referenced in the bibliography.

(5) OPEN-PLAN LAYOUT – ESCAPE WINDOWS

In determination reference 45/1/178 (22 June 1999) the Secretary of State did not consider that a first floor escape window would compensate for the absence of a protected stair discharging to a final exit in the case of a loft conversion forming a new storey in an existing two-storey house. It should be noted that the paragraph references in the following excerpt refer to an earlier version of Approved Document B. However, the principle holds for the current version (2000 as amended 2002).

The Department accepts that you have provided fire separation between the new loft room and the lower rooms but it does not consider that you have provided an alternative escape route leading to its own final exit, as is suggested in paragraph 1.20(b). In your case the existing stair is shown to discharge into the ground floor lounge with access to a final exit door being available only through the kitchen.

The Department does not consider that the provision of an escape window on the first floor is an acceptable compensatory feature in lieu of the protected stair discharging to a final exit in accordance with either Diagram 2 (a) or (b). Neither does the Department consider that an escape window is an acceptable compensatory feature to an alternative escape route leading to its own final exit if the intention is, as you have stated, to follow paragraph 1.20(b).

The definition of a final exit given in appendix E of the Approved Document suggests that it should be sited to ensure the rapid dispersal of persons from the vicinity of the building and the definition does not imply that a window is suitable as a final exit. To reinforce this view, paragraph 1.29 of the Approved Document suggests that a window at first floor level is suitable for self-rescue; the paragraph does not suggest that a window is also suitable for the rapid dispersal of persons.'

(6) DEFINITION OF A STAIR

It is sometimes difficult to achieve adequate headroom in loft conversions. Whilst there is no guidance on minimum floor to ceiling height within rooms, Approved Document K *Protection from falling, collision and impact* sets out minimum clearances for stairs and landings. Whilst these are unequivocal, defining exactly what constitutes a 'stair' for the purposes of the guidance is less clear, particularly in conversions where there are relatively complex configurations of steps and landings.

This is the subject that led to two appeals against refusal to relax requirement K1, both dated 6 February 2004 (references 45/3/165 and 45/3/161). The outcome in both cases was the same. The text of appeal 45/3/165 is reproduced below.

The appeal

This appeal relates to completed building work to create a 31.5 m² bedroom with integral en suite shower and WC in the roof space of a four-bedroom (previously two-storey) detached house approximately 8 m × 6 m in area (i.e. a loft conversion). The roof is of pitched, single ridge construction running between the flank walls; and the new second floor room has been created by breaking open approximately three-quarters of the length of the rear pitch from eaves level and constructing a dormer framework containing three separate windows, the centre one of which is designated as an escape window.

Access to the new room is by a timber stair installed over the ground to first floor stair. At the foot of the new stair are two winders and at the top there is a quarter 'drop landing' giving access via an additional step, facing the stair, to the new room. The room is protected by an inward opening fire door located on the top of the additional step.

These proposals were the subject of a full plans application which was conditionally approved. This included a condition to ensure a 2 m minimum headroom at the head of the stair. However, it is understood that the floor and dormer roof were not built to the levels indicated on the approved plan with the result that the Borough Council considers that the headroom for the stair at the top (i.e. additional) step is not in compliance with Requirement K1.

However, you took the view that Requirement K1 should not be applied to the question of headroom from this additional step up from the 'drop landing' into the new room. You therefore applied to the Borough Council for a relaxation of Requirement K1 which was refused. The Council then issued a notice of contravention in respect of Requirement K1 requiring corrective works within 28 days. It is against the refusal to relax Requirement K1 that you appealed to the Secretary of State.

The appellant's case

You consider that the stair to the new second floor room comprises the main flight and the drop landing, and that you have provided approximately 2 m headroom throughout. You argue that the headroom from the additional step at the door into the new room should not be taken into consideration because it is not part of the flight.

You also state that you have discussed the form of construction with an adjacent Borough Council's building control division who consider that your proposals comply with the current Building Regulations.

The Borough Council's case

The Borough Council has considered the definition of a 'stair' given in Approved Document K *Protection from falling, collision and impact*, which is: 'a succession of steps and landings that makes it possible to pass on foot to other levels'. The Council regards the landing, the additional step at the doorway to the new second floor room, and the doorway itself as forming part of the stair. The Council points out that the headroom under the doorway is approximately 1.76 m and is approximately 1.86 m under the ridge in the new room. Given that Approved Document K recommends that 2 m is adequate headroom, the Council does not consider that Requirement K1 has been complied with.

The Borough Council also considers that the door sweeping across what it regards as the upper landing contravenes the guidance in paragraph 1.16 of Approved Document K, which says that: 'To afford safe passage landings should be clear of permanent obstruction.'

To support its case the Borough Council has enclosed a copy of an appeal decision by the Secretary of State in a case involving headroom which he issued in 1996.

The Secretary of State's consideration

Falls on stairs in dwellings are a very common type of accident resulting in about 500 deaths per year and many thousands of injuries. The Secretary of State therefore considers that good stair design makes an essential contribution to life safety.

In considering this appeal the Secretary of State has first considered to what degree the proposed stair may fall short of compliance with Requirement K1, thereby potentially warranting a relaxation of this requirement.

Requirement K1 says that: 'Stairs, ladders and ramps shall be so designed, constructed and installed as to be safe for people moving between different levels in or about the building.' The guidance in Approved Document K gives solutions for common situations, but loft conversions often present particular problems which have to be considered on their individual merits. The overriding consideration is the safety of the stair user.

For the purposes of Part K *Protection from falling, collision and impact*, a stair is defined as: 'A succession of steps and landings that makes it possible to pass on foot to other levels.' You and the Borough Council disagree as to whether the stair ends at the foot of the additional step and new door, or continues into the proposed new second floor room. In the Secretary of State's view, if there was no door, then the stair should be considered to include the single additional step together with the landing beyond formed by the adjacent part of the floor of the new room. However, in this case there is a door, and the Secretary of State considers that it is reasonable to regard the stair as ending at that door, even though this landing is not strictly at the upper level. This is because the door provides a clear barrier to progress, which requires the user to stop and open it before proceeding. The door also provides a clear marker for the change in level at the single additional step.

The guidance on performance on page 5 of Approved Document K makes it clear that Requirement K1 will only be applicable to differences in level of more than 600 mm. Given that it is the Secretary of State's view that the stair ends at the additional step and the door, it follows that the step is not required to comply with Requirement K1 of the Building Regulations because the difference in levels here is less than 600 mm. It also follows that because the floor area of this part of the new room cannot be considered to be a landing, the room height of this part of the room is not subject to the Building Regulations.

The Secretary of State has also noted the Borough Council's reference to a previous appeal decision made in 1996, which the Council contends supports its case. However, the Secretary of State is required to consider all cases on their individual merits. He considers that loft conversion cases can pose different questions, and issues specific to previous cases will not necessarily be relevant to subsequent ones.

In a more general context than the precise application of the Building Regulations, the Secretary of State recognises that the overall concern of the Borough Council is the safety of those using the stair. Although there is adequate headroom to the stair itself, the Council's concern is with the limited headroom of 1.76 m under the doorway and approximately 1.86 m under the ridge in the new room. These figures are in contrast to the figure of 2 m which is defined as adequate headroom on the access between levels in paragraph 1.10 of Approved Document K. However, paragraph 1.10 recognises the constraints which may exist in loft conversions and says: 'For loft conversions where there is not enough space to achieve this height, the headroom will be satisfactory if the height measured at the centre of the stair width is 1.9 m reducing to 1.8 m at the side . . .'

Although the Borough Council's concerns over the limited headroom relate to construction elements which in the Secretary of State's view do not fall to be controlled under the Building Regulations (i.e. room height), he does consider it appropriate to comment as follows. As a general principle he takes the view that where it is necessary to use a drop landing, an additional step – or additional step and door – should preferably be at right angles to the direction of the main stair flight, and that the length of the drop landing should be of sufficient length to enable the 90° turn to ascend the step to be made some distance away from the top of the main flight. Such a design principle should minimise any hazard and risk of falling. When ascending the additional step any person who did bump their head is likely to do so in a position of 90° to the flight, thus increasing the chances of regaining their balance on the drop landing as opposed to falling back down the stair flight.

Notwithstanding the above general comments regarding optimum design for life safety, the Secretary of State does accept that loft conversions can present constraints on stair design, particularly in terms of headroom. As noted above, this is acknowledged in Approved Document K. Thus although the Secretary of State considers that there may be potential to further improve the design safety of this particular stair, he considers that the risk of harm to the users of the stair is acceptably small, given that they will generally be familiar with the layout. He therefore takes the view that the stair as constructed offers a reasonable level of safety and therefore complies with Requirement K1. It follows that he considers it would be neither appropriate nor necessary to relax Requirement K1 in order to secure the compliance of the existing stair.

The Secretary of State's decision

The Secretary of State has given careful consideration to the facts of this case and the arguments put forward by both parties. Paragraphs 12–20 above have considered and given the Secretary of State's view on the compliance of the stair as presently installed, having regard to the circumstances of this particular case.

However, you have appealed to the Secretary of State in respect of the refusal by the Borough Council to relax Requirement K1. The Secretary of State considers that compliance with Requirement K1 makes an essential contribution to life safety and as such he would not normally consider it appropriate to relax it, except in exceptional circumstances. Moreover, because in the particular circumstances of this case he considers that your building work complies with Requirement K1, there would appear to be no prima facie case to relax the requirement in any event. Therefore, taking these factors into account, the Secretary of State has concluded that it would not be appropriate to relax Requirement K1 *Stairs, ladders and ramps* of Schedule 1 to the Building Regulations 2000 (as amended). Accordingly he dismisses your appeal.

(7) STAIR DESIGN – PITCH AND GOING

Stair designs that diverge from established patterns are potentially problematic and this is borne out in the following two appeals. In an appeal against refusal to relax Requirement K1 dated 4 February 1999 (reference 45/3/130) a Borough Council had served a notice under section 36 of the Building Act 1984 to remove or alter a staircase in order for it to comply with the Building Regulations.

The appellant's staircase comprised four winders at both top and bottom with an intermediate straight section. The Borough Council considered the stair as a unit to be unsatisfactory because, in its view, the pitch exceeded 42 degrees, the going varied from step to step and the going of some of the steps was considered to be inadequate.

The appeal was dismissed on the grounds that Requirement K1 can be a matter of life safety. However, the Secretary of State took the view that this particular stairway would comply with the requirement if modified:

The appropriate design and measurements for winder stairs are covered in BS 585: 'Wood stairs' Part 1: 1989. This recommends that the going of the straight part of the flight should be as recommended in BS 5395 ('Stairs, ladders and walkways' Part 1: 1977 (confirmed November 1984) 'Code of practice for the design of straight stairs'), and that the centre going of the winders should be uniform, and not less than the going of the straight part of the flight. The smallest going recommended in BS 5395: Part 1 is 225 mm. However, Part 2 of that BS deals with spiral and helical stairs and allows centre goings as small as 145 mm in situations such as this.

Clearly, your stair falls well short of the guidance in BS 585, but it is within the limits for spiral stairs. Small spiral stairs are considered to be safer than straight or winder stairs because, in the event of a fall, the user is likely to fall towards the guarding/handrail, thus providing them with an opportunity to regain balance.

Although not shown in the photographs you have submitted, you state that additional features have been added to the stair including extra handrails, grab rails and handles. This would improve the safety, especially if the handrails are continuous.

In loft conversions such as this, the guidance in Approved Document K takes account of the limited space available and the light use – mainly by people who are familiar with the stair – and suggests provision of either a fixed ladder or an alternating-tread stair. In the Department's view the stair as proposed and installed would be as safe as these alternatives and in the context of this particular situation complies with Requirement K1.

(8) STAIR DESIGN – SELECTIVE APPLICATION OF REGULATIONS

The risks posed by stairs with open risers and without guarding is considered in an appeal against refusal to dispense with Requirement K1 (reference 45/3/141 dated 13 January 2000). The appeal concerned stair access for a bungalow loft conversion. The stair had been designed for aesthetic effect with open risers of approximately 200 mm. A handrail was provided on one side but the stair was open on the other. The appellant was prepared to change the stair to comply with the requirements of the Building Regulations were the house to be sold.

The District Council expressed the view that the requirements of the Building Regulations

'cannot be taken in part for the particular occupants at the time to choose the elements which are felt to apply to them.'

This view was supported by the Secretary of State:

The Department notes and endorses the point made by the District Council that the Building Regulations cannot be applied selectively but must relate to all persons, including young children, in or about a building; and that they must be applied at the time of the building work. In the Department's view, compliance could be readily achieved in practical terms in respect of both the guarding and the open risers. Moreover, the effect of the open risers need not necessarily be completely compromised if an adequate guarding were to be installed to ensure that a 100 mm diameter sphere could not pass through the risers, as recommended in paragraph 1.9 of Approved Document K.

(9) MEANS OF ESCAPE – FLOORS MORE THAN 7.5 m ABOVE GROUND LEVEL

At the time of writing, the only definitive guidance on escape from floors that are more than 7.5 m above ground level is that an alternative means of escape be provided. This recommendation is contained in BS 5588 and in Approved Document B *Fire safety*.

However, it is often not practical to retrofit an alternative means of escape in a relatively narrow urban terraced dwelling when a loft is converted: there is rarely enough space for an additional internal stair, and planning regulations generally preclude the construction of an external fire escape.

The basis for providing appropriate means of warning and escape in one particular case is described in determination 45/1/200 (29 April 2002), which is reproduced in full here. A similar approach is described in Chapter 4.

The proposed work

The proposed building work comprises a loft conversion to a three-storey mid-terrace, four bedroom town house of approximately 45 m² in plan area. The conversion will form a single room of 27 m² in the roof space.

Two pairs of roof rescue windows are to be installed (one above the other) in both the front and rear roof slopes. The lower window of each pair will be 780 × 1400 mm and will be located at a maximum distance of 1700 mm along the roof slope from the gutter board.

As existing the ground floor of the house comprises a garage and hall to the front, and a kitchen/dining room to the rear. The first floor accommodates bedroom no. 1 (with an en suite bathroom) and a sitting room with sliding door opening onto a balcony formed above the front protrusion of the garage, porch and meter cupboard below, and which has a depth of 2 m. The second floor accommodates three bedrooms and a separate bathroom to the rear and left on plan.

The stairs to each floor are located in the centre of the house and transverse to the party walls. They rise in single straight flights, one above the other, to landings on the first and second floors – the whole stairway being protected to a standard of 30 minutes of fire resistance. In addition the doors to the kitchen/dining room and the garage are fitted with smoke seals. The new stair to the proposed third floor (fourth storey) will be achieved by 180° winder stairs at the second floor landing level, rising thereafter in a straight flight. This new stair will be protected at second floor landing level by a 30-minute fire-resisting door to be installed 900 mm from the base of the new stair thus separating the new stair and the door to the second floor bathroom directly adjacent to the foot of the stair, from the rest of the second floor landing.

A mains operated automatic fire detection and alarm system is to be installed and will comprise detectors in all rooms (except the ground floor cloakroom and the first floor en suite bathroom); the first and second floor landings; and above the proposed new stair. Heat detectors will be installed in the garage and within the cooking area of the kitchen/dining room with a smoke detector provided in the dining area itself. All the other detectors will be smoke detectors.

From the maps and block plan provided, it is apparent that the front elevation of the house faces east south east with approximately 300 m of public open space. The rear garden backs onto an estate road.

These proposals were contained in a full plans application which was rejected by the Borough Council on the grounds of non-compliance with Requirement B1 of the Building Regulations. The Council was concerned, in particular, that your proposals would result in a new storey 7.9 m above ground level from which assisted escape might not be practical. However, you consider that the proposed fire protection to the single stairway, and the inter-linked fire detection and alarm system, provide adequate compensation to allow for the omission of an alternative means of escape from the new storey whilst still achieving compliance with Requirement B1. It is in respect of this question that you have applied for a determination.

The applicant's case

You take the view that the Borough Council's requirement for an alternative means of escape could only be provided by an external stair which would be too onerous for the provision of a single loft room. Furthermore, you believe that it is unlikely that planning permission would be granted for such a stair.

You have referred to a number of previous determinations by the Secretary of State which you consider support your case on the grounds that they seem to accept the principle of a suitable mains operated automatic fire detection and alarm system (incorporating interlinked detectors fitted throughout the building) as a substitute for an alternative means of escape.

You also refer to the following provisions contained within your proposals:

(i) The internal stair is to be fully protected, enclosed in fire-resisting construction with fire-resisting self-closing doors.

(ii) Interlinked smoke alarms are to be provided in all habitable rooms including the dining area of the kitchen/dining room, and circulation spaces. Heat detectors will also be installed within the cooking area in the kitchen/dining room and the garage.

(iii) A roof rescue window is to be installed in each of the front and rear roof slopes in accordance with Diagram 6 (b) of Approved Document B *Fire safety*. Although you accept that the Approved Document indicates that this diagram refers to two-storey houses, you consider that it is conceivable that the occupants of the loft room in your case could be rescued by the Fire Service at this height.

(iv) You also accept that Clause 4.4 (d) of BS 5588: 'Fire precautions in the design, construction and use of buildings: Part 1: 1990 Code of Practice for residential buildings' recommends that for dwellings with a storey situated at 7.5 m above ground level an alternative escape should be provided. However, you point out that your drawings indicate that the underside of the proposed new third floor is only 7.7 m above ground level and the top of the new floor is no more than 7.9 m above ground level.

In response to some of the Borough Council's concerns you have also added that:

(i) It is because of the ground floor level projection at the front of the building possibly impeding rescue ladders, that a second roof rescue window is to be installed in the rear roof slope at third floor level. You add that a ladder can be erected at the rear of the house where at ground floor level there is a level, paved rear garden having direct and level access to a public highway.

(ii) Your proposed mains operated automatic smoke and heat detection system is intended to give early warning in the event of fire. You also indicate that Approved Document B, in discussing the provision of emergency egress window and external doors, does not preclude the provision of these in rear elevations where there is a direct escape route from the back garden to reach a place free from danger from fire. In discussing rescue by ladder the Approved Document gives no guidance as to the internal arrangement of rooms comprising the layout of the dwelling.

(iii) As well as the installation of the alarm system additional fire precautions are proposed on the lower floors in order to make the single escape stair as safe as possible.

You conclude that you believe that the Secretary of State has to consider each case on its merits and is therefore unlikely to be able to take account of other similar properties in the area.

The Borough Council's case

The Borough Council has raised the following concerns relating to your proposals:

(i)   In considering your application the Borough Council has taken the view that Requirement B1 is a life safety matter which is directly related to the health and safety of occupants and visitors to the building and anyone who may try to rescue people in the event of fire.

(ii)   Your proposals will result in there being two storeys situated at a height greater than 4.5 m above ground level and which therefore do not accord with either paragraphs 2.12–2.14 of Approved Document B or the alternative guidance contained within paragraph 4.4(d) of BS 5588: Part 1. In the Borough Council's view, you are therefore required to demonstrate that your alternative proposals are no less safe for means of escape in case of fire than would be achieved by following one of these standards.

(iii)   Although you refer to the intended use of the loft room as a study in your application for a determination, your full plans submission refers to the conversion as forming a habitable room. As such, the room could be used for sleeping accommodation which would heighten the risks associated with escape in the event of fire. The Building Regulations do not provide for ongoing control of the use of part of a building.

(iv)   The exposed location of the house brings into question whether rescue using either of the two roof rescue windows on the third floor could be achieved on anything other than a calm day and whether this provision therefore represents an effective alternative escape/rescue option.

(v)   The ground floor level projection of the building by the garage, meter cupboard and porch, impedes rescue ladder access to the front slope of the roof. Rescue from this elevation could only be effected if a ladder were erected outside of its safe limits for use, which would place rescuers at risk.

(vi)   Given the impediments associated with rescue from the front elevation of the building, alternative rescue and escape could only be effected from the rear elevation. However, build-up of fire in any of the rear rooms within the house – including the kitchen where statistically the highest incidence of fire occurs – could preclude the use of the rear rescue window.

(vii)   With reference to the above points, the Borough Council takes the view that you appear to be placing heavy reliance on early warning of a fire and the single escape stairway remaining free of fire for a long enough period to achieve escape. The Fire Authority does not support this approach.

The Borough Council also questions your contention that the previous determinations you refer to support your case and concludes by stating that there are a number of residents of other similar properties in the area that may also be interested in a loft conversion and the decision in your case may set a precedent for others to follow.

The Secretary of State's consideration

In the Secretary of State's view, the main consideration in this case is the safety of the occupants of the new third floor if a fire occurs at a lower level. The Borough Council has suggested that because this new floor is more than 7.5 m above ground level an alternative escape route from this storey should be provided. You consider that the provision of an alternative escape route would be too onerous a provision for a single loft room and that it is highly unlikely that planning permission would be granted for such a stair. Instead you have proposed a package of features intended to compensate for the omission of the additional escape route. This package includes a protected stairway, interlinked smoke/heat alarms provided in all habitable rooms, fire-resisting separation of the new top floor, and the provision of front and rear roof rescue windows for assisted escape.

The Secretary of State considers that additional measures, such as those described in Approved Document B, are necessary for floors in houses more than 7.5 m above ground level to address the increased risk of the occupants of a floor becoming trapped at this level. The increased risk is due to the additional time it will take to travel down the stairway and the reluctance of the occupants to use an escape route which may be becoming obscured by smoke.

The Borough Council has expressed concern that your proposal to provide windows for assisted escape from the new loft room might not be practical. The approach provided in Approved Document B for loft conversions is a departure from the general principle that escape should be provided without outside assistance; and you have acknowledged this position. The concession in respect of loft conversions is only intended to be used where the roof space of an existing two-storey house is converted. The use of roof rescue windows for assisted escape for loft conversions of above this height is not considered appropriate, albeit that it may be possible in practice. It follows that whilst the Secretary of State does not completely discount in this particular instance the relevance of making provision for roof rescue windows at third floor level, he has not considered it appropriate to regard such provision as constituting an acceptable alternative means of escape for the purposes of determining compliance with Requirement B1.

However, the Secretary of State does recognise in this particular case that your proposed provision for an enhanced level of early warning comprising the use of interlinked smoke/heat alarms in each habitable room, in conjunction with a fully protected primary escape route – including 30-minute fire-resisting and self-closing doors – will reduce the risk of the occupants of the new storey becoming trapped. This level of provision needs to be compared with the level of safety which will be afforded the occupants of a typical three-storey house where smoke alarm provision may have been limited to the stairway only, in accordance with the guidance given in paragraphs 1.2–1.22 of Approved Document B. In the Secretary of State's view, the level of safety to be provided by your proposals for the occupants of the proposed loft room will be of a similar level and will therefore achieve compliance with Requirement B1.

The determination

The Secretary of State has given careful consideration to the particular circumstances of this case and the arguments presented by both parties. He has also noted that you have referred to previous determination decisions which you contend support your case. However, the Secretary of State is required to consider all cases on their individual merits and some issues which are specific to previous cases will not be relevant to subsequent ones.

As indicated above, on the basis of your proposals as submitted the Secretary of State considers that they make adequate provision for warning and safe escape. He has therefore concluded and hereby determines that your proposals comply with Requirement B1 *Means of warning and escape* of Schedule 1 to the Building Regulations 2000 (as amended).

(10) MEANS OF ESCAPE – LADDER BRACKETS

In certain circumstances where a loft conversion forms a new third storey (described in Chapter 4), there is no need for a *fully* protected stair provided that an emergency egress window is provided to allow ladder-assisted rescue. However, the position of such a window relative to practical obstacles that would impede the use of a ladder must be considered.

In determination 45/1/211 (1 July 2004), the applicant proposed the provision of an escape window situated over a narrow (1.1 m wide) passage. Because a ladder erected in such a position would be too steep to climb safely, the proposal included the installation of a welded mild steel ladder bracket or 'keep' that would allow a ladder to be set up at right angles to the escape window.

A considerable number of objections to the proposal were raised by the Borough Council and Fire Authority. These included concerns about the condition and fixings of the ladder bracket (which would not be known to anybody attempting a rescue) and, indeed, the ability of any rescuer to identify what the bracket was for.

The Secretary of State considered that such brackets should only be used 'in cases where there is no alternative available to facilitate escape; and where the bracket is robustly designed and fabricated, and is securely fixed.' The requirements for maintenance and durability of the bracket were also considered. It was noted that these points 'could be addressed by ensuring that there is sufficient redundancy in the fixing design and by using materials, such as stainless steel, that are inherently resistant to corrosion.' The Secretary of State determined that the proposals did not comply with Requirement B1.

(11) MEANS OF ESCAPE – STRUCTURAL CONSTRAINTS

The subject of an appeal against a refusal to relax requirement B1 *Means of escape*, reference 45/3/143 (8 May 2000) was the position of a fire escape window for a loft conversion that formed a new third storey. The appeal related to work that had been completed.

In this case, an escape window of appropriate dimensions was set in the rear roof slope with access stated to be available in the rear garden. However, the bottom of the window was 2.2 m from the eaves, rather than the 1.7 m recommended in Approved Document B *Fire safety*. The applicant indicated that this was because of structural constraints in the arrangement of the roofing timbers.

The Borough Council contended that a rear extension below this window formed an impediment to ladder access (note: guidance on the effect of ground-storey extensions on ladder rescue had not been included in Approved Document B at the time of the appeal).

The Secretary of State concluded that:

> . . . in this case the location of the escape window from the new second floor is unsatisfactory with respect to both the excessive distance of the window sill from the eaves and in respect of the rear ground floor extension which has the potential to make safe escape more difficult. He has concluded that the structural constraints which apparently determined the location of the escape window do not amount to extenuating circumstances such as would justify relaxing Requirement B1 *Means of escape* . . . and that the Borough Council therefore came to the correct decision in refusing to relax this requirement. Accordingly, he dismisses your appeal.

(12) SLEEPING GALLERY

Under current guidance (Approved Document B 2000, amended 2002), a proposal to create a gallery in the circumstances set out below would not normally be acceptable. In this case, however, the gallery and room below it were so closely linked that independent use of the gallery was considered unlikely. The Secretary of State thus determined that the proposal complied with Requirement B1. The full text of the determination letter (1 July 2004, reference 45/1/213) is set out below:

The proposed work

The building work which is the subject of this determination is a first floor extension to a two-storey, double fronted, three bedroom detached house whose plan footprint is 'L' shaped – being approximately 10 m in frontage and 10 m in depth. The site slopes from front to rear.

The ground floor is of split level design and comprises a hall, lounge and sitting room at the front with a kitchen to the centre rear. Three steps from the kitchen lead down to a single-storey extension comprising a dining room and a utility room located on the side and rear of the house.

The first floor comprises three bedrooms to the front with a bathroom to the rear located above the kitchen. There is a drop landing at the top of the stairs with an additional step up to each of the two landings and accommodation on either side of the stair.

The proposed alterations and the extension are at first floor level and will involve building above the existing single-storey extension to provide an additional bedroom and en suite bathroom. The construction will incorporate a gable end to the rear elevation. Two inward opening French windows are proposed in the new rear elevation of the bedroom which give access to a narrow width balcony.

In addition, a gallery of approximately 2 m width is proposed which will run across the full width of the extension (approximately 6 m) – i.e. running across the rear of the bedroom and across the total area of the en suite bathroom below. The gallery will be open to the remaining area of the bedroom. An oriel window to the gallery in the gable end will provide the minimum dimensions required for an escape window. The height of the gallery floor above the bedroom floor is approximately 2.4 m. The plans indicate that the height of the gallery floor above ground level at the rear is approximately 5.2 m.

Access from the existing first floor landing to the new bedroom and gallery is from a new self-closing 30-minute fire door which opens onto a new landing within the bedroom. This landing gives access to the right of the door to three winder steps leading down to the new bedroom floor; and directly opposite the door to a stair which leads up to the gallery. The maximum travel distance from any point on the gallery to the head of this stair is 4 m.

All existing internal doors opening off the hall and the two first floor landings are shown on plan as self-closing. Interconnected mains operated smoke alarms are proposed in the hall at ground floor level; the first floor drop landing; and at the top of the stair to the gallery.

These proposals formed the basis of a full plans application which was rejected by the Borough Council on the grounds, inter alia, of non-compliance with Requirement B1 in respect of safe escape from the gallery level. The Council considered that the proposed gallery was more akin to the conversion of a roof space in a two-storey house and as such contravened Requirement B1. In the Council's view to achieve compliance it would be necessary to separate the gallery from the rest of the house using fire-resisting construction. However, you have argued on various grounds that your proposals do comply with Requirement B1 and it is in respect of this question that you have applied to the Secretary of State for a determination.

The applicant's case

You have made the following points to support your case for compliance with Requirement B1:

(i)   The gallery area is to be used in conjunction with the bedroom, but not as sleeping accommodation.
(ii)   Because of the split level layout of the house the gallery is not a full storey height above the entrance level to the bedroom, but is at split level with the bedroom.
(iii) The gallery is part of the actual bedroom and not a separate room. Should there be a fire or another emergency, the escape distance from the gallery is less than from the far end of the main part of the bedroom. You also consider that, in such a case, it would be more dangerous if the gallery were to be enclosed.
(iv)   With reference to paragraphs 2.12 to 2.26 of Approved Document B *Fire safety* regarding provisions for houses with a floor more than 4.5 m above ground level, you acknowledge that the land on which the house is built is not level but slopes down from front to rear, thus resulting in the gallery floor at the rear being more than 4.5 m above ground level. However, you argue that, as the building has been built to a split level design, it lends itself to a mezzanine/gallery type extension rather than to the creation of a very tall room. On this basis you therefore emphasise that the gallery is an integral feature of the bedroom and not a main floor.

The Borough Council's case

The Borough Council has made the following points to support its case for rejecting your proposals:

(i)   The gallery area is part of a bedroom and as such could easily be used for sleeping accommodation. Paragraph 2.9 of Approved Document B states that where a sleeping gallery is provided, the gallery should be no more than 4.5 m above ground level. Even taking into account the fact that the bedroom below is to have a ceiling height of no more than around 2 m, the proposed floor level of the gallery will still exceed the 4.5 m threshold. In the Borough Council's view your proposals do not therefore accord with the guidance in the Approved Document.

(ii) In the Borough Council's view the proposed gallery could be treated on the same basis as a loft conversion. As such this would require the gallery to be separated from the rest of the house by fire-resisting construction.

(iii) The Borough Council countered your points about what the position might be if the land was level by stating that if the site were level your property would not have been constructed as a split level house. It reiterates the point that the height of the gallery floor exceeds 4.5 m from ground level, as detailed in Diagram C5 (page 129) of Approved Document B – Height of top storey in building.

The Secretary of State's consideration

The Secretary of State notes that in this case a new split level bedroom (i.e. bedroom plus gallery) is proposed to be introduced into an existing two-storey house. This will, in effect, create an additional storey which is more than 4.5 m above ground level when measured from the lowest ground level. At this height it is not considered to be safe for people to make their own escape from windows. It would normally be necessary, therefore, to provide a protected escape route down through the house formed with fire-resisting construction and fire-resisting self-closing doors.

Paragraphs 2.17–2.26 of Approved Document B provide guidance for loft conversions. This is based on the strategy of providing fire-resisting construction to separate the new floor from the rest of the house, and thus allow the occupants of the new floor to wait in relative safety for rescue via a ladder through a suitably placed window. Using this approach existing doors should be made self-closing but need not always be replaced.

In this particular case, in order to take advantage of the split level layout of the house, you have proposed that the new floor will be open to the room below in the form of a gallery. The Borough Council has taken the view that this would not be acceptable but has suggested that the new floor could be treated as a loft conversion. Whilst as noted above, this would not necessitate the replacement of the doors to the stairway elsewhere in the house, it would result in the need to separate the new storey (i.e. the gallery) from the rest of the house, including the bedroom below. It would then constitute a small loft room rather than a gallery.

You consider that the gallery itself is very small and the vertical distance from the gallery floor to the first floor landing is less than a conventional storey height. The implications of your comments appear to be that the safety of people using the gallery would be no worse than that for the people in the bedroom underneath and that therefore no additional fire protection is required.

The Secretary of State takes the view that gallery floors can be similar to inner rooms. This is because there is an increased risk that persons on the gallery floor may become trapped should a fire occur in the room through which it is accessed. Where a gallery is below 4.5 m from ground level it is possible to mitigate this risk by the provision of an alternative escape route such as a window or by ensuring that travel distances in the gallery are limited.

You have stated that the gallery area is to be used in conjunction with the bedroom, but not as sleeping accommodation. However the Borough Council has argued that the gallery area is part of a bedroom and as such could easily be used for sleeping accommodation. The Secretary of State takes the view that the gallery, as with any habitable room, may well be used for sleeping and this should be a material consideration.

The Secretary of State recognises that, in this case, the gallery floor is more than 4.5 m above ground level. However, taking into account the small size of both the gallery and the bedroom below and their inter-relationship to each other, the two levels could only conceivably be used in conjunction with one another. This position must be contrasted with a separate loft room which may be put to different or ancillary purposes. Therefore, given this interrelationship between the gallery and the bedroom, and the provision of a smoke alarm above the gallery stair, it would be reasonable to treat both levels as being part of the same new first floor bedroom.

The Secretary of State has therefore concluded that additional protection for the occupants of the gallery above and beyond what would be provided for the bedroom, is not necessary. In this case an escape window has been provided at first floor level (i.e. the French windows) which could be utilised by the occupants of the bedroom and gallery should they become trapped by a fire elsewhere in the house. As such the plans achieve compliance with Requirement B1.

The determination

The Secretary of State has given careful consideration to the particular circumstances of this case and the arguments presented by both parties.

As indicated above, the Secretary of State considers that your proposals as submitted, which incorporate an open gallery to the new bedroom, make adequate provision for warning and escape. He has therefore concluded and hereby determines that your proposals comply with Requirement B1 *Means of warning and escape* of Schedule 1 to the Building Regulations 2000 (as amended).

Appendix 3 Planning and curtilage

There is no statutory definition of the term curtilage, and local authorities interpret its meaning differently. Some now require planning permission for raising a party wall at full thickness on the grounds that half the wall is outside the curtilage of the building being extended. Others consider that the curtilage includes the whole thickness of the party wall and issue Certificates of Lawfulness confirming that such development is permitted.

Reasons for *excluding* the raising of a party wall from permitted development were advanced in a 1997 appeal concerning a full-width conversion in a mid-terrace house in Bath (reference APP/C/96/F0114/642257). An apparent consequence of this is that an increasing number of local authorities now require an application for planning permission for such work, and a Certificate of Lawfulness (which confirms that a proposal does not require express planning permission) is refused.

In at least one case (London Borough of Merton, 2005) a local authority has granted a lawful development certificate for the erection of *half* a party wall in the case of a roof extension.

Arguments for *including* a party wall within the curtilage of a dwellinghouse in the context of a roof extension are set out in a planning inspector's decision concerning a house in the London Borough of Enfield. This decision letter is reproduced in full below. It should be noted that a planning inspector's decision is definitive only in so far as it applies to a particular property at a specific point in time – it is not case law.

Appeal Ref: APP/Q5300/X/01/1062324
[-----] *Road, Enfield, Middlesex*

- The appeal is made under section 195 of the Town and Country Planning Act 1990 as amended by the Planning and Compensation Act 1991 against a refusal to grant a lawful development certificate (LDC).
- The appeal is made by Mr [-----] against the decision of London Borough of Enfield.
- The application (Ref. LDC/00/0300), dated 30 October 2000, was refused by the Council by notice dated 15 January 2001.
- The application was made under section 192(1)(b) of the 1990 Act as amended.
- The development for which a Certificate of Lawfulness is sought is loft conversion of L-shaped roof into L-shaped rear facing dormer loft, including raising of lower part of L-shaped roof party wall, to contain 2 × bedrooms, 1 × bathroom and boiler room.

Summary of decision: The appeal is allowed and a Certificate of Lawfulness is issued, in the terms set out in the formal decision below.

Preliminary matters

(1) The appeal property is a two-storey, brick built semi-detached house on the southern side of [-----] Road. In common with the adjoining house at No [--], it has a back addition. The proposal includes the raising in height of the party wall to this addition by some 0.5 m. The Council's only reason for refusal of the certificate is based on their belief that the extension to the party wall would not fall within Class B of Part 1 of Schedule 2 to the Town and Country Planning (General Permitted Development) Order 1995 (GPDO) by reason of part of it being outside the curtilage of the dwellinghouse.

(2) There is no dispute that the party wall straddles the boundary between the two properties. The submitted plans show the upward extension of it to be of the same width, so that it too would straddle the boundary. The Appellant now argues however that his proposal is made only in relation to that part of the structure within his property, with his neighbour able to exercise similar permitted development rights in relation to the adjoining property. It is thus necessary for me to determine, as a preliminary issue, the nature and extent of the proposal before me.

(3) The plans do not show the precise method of construction intended for the additional courses. Following the present pattern, however, and given the nature of the structure, it is reasonable to expect that some bricks would be set parallel to and others at right angles across the boundary. Even if that were not so, courses set either side of the boundary would most probably need to be bonded in some way, and capped as shown on the submitted plans. The wall can therefore only sensibly be regarded as a single structure, and the adding of extra courses to it as a single operation or development.

(4) It follows that, even if there were applications from both owners, each would amount to a separate scheme involving the same development, not as two schemes with each affecting only their own property. The fact that the adjoining owner may have given permission to the Appellant to carry out the development, whether pursuant to The Party Wall Act or otherwise, does not alter the position for planning purposes, any more than does the grant of approval under the relevant Building Regulations.

Main issue

(5) Part 1 of Schedule 2 to the GPDO carries the heading 'Development within the curtilage of a dwellinghouse'. For permission to be granted by its provisions therefore, the development must be within that curtilage. In the light also of my conclusion on the preliminary issue, the main issue is thus whether the extension of the party wall would meet that description, despite the fact that it would straddle the boundary.

Inspector's reasons

(6) As the Council point out, there is no statutory definition of the term 'curtilage', but it has been considered by the courts on a number of occasions. It was said in *Methuen-Campbell* v. *Walters [1979] 1 QB 525 (CA)* that for one piece of land or building to fall within the curtilage of another, the former must be so intimately associated with the latter as to lead to the conclusion that the former in truth forms part and parcel of

the latter. In *Dyer* v. *Dorset County Council [1988] 3 W.L.R. 213*, the Court of Appeal held the term to carry its restricted and established meaning connoting a small area forming part or parcel with the house or building which it contained or to which it was attached. The definition in the Oxford English Dictionary was said to be 'adequate for most present day purposes'. This described it as a 'small court, yard, garth or piece of ground attached to a dwellinghouse, and forming one enclosure with it, or so regarded by the law; the area attached to and containing a dwelling-house and its outbuildings'.

(7) In *Attorney-General ex. rel. Sutcliffe, Rouse and Hughes* v. *Calderdale Borough Council [1983] JPL*, the Court of Appeal accepted that three factors had to be taken into account in determining what constituted the curtilage, namely: (1) the physical layout of the building and structure; (2) their ownership, past and present; and (3) their use and function, past and present. In a later case, *McAlpine* v. *Secretary of State for the Environment [1995] JPL B43*, the High Court identified three relevant character-istics of curtilage. First, it was confined to a small area about a building; second, an intimate association with land which was undoubtedly within the curtilage was required in order to make the land under consideration part and parcel of that undoubted curtilage land; and third, it was not necessary for there to be physical enclosure of that land which was within the curtilage, but the land in question at least needed to be regarded in law as part of one enclosure with the house.

(8) The Council's case is dependent on the proposition that land beyond the ownership boundary is necessarily outside the curtilage, but they cite no authority in support of this. It is hard however to imagine a structure, and thus the land on which it stands, having a more intimate association with the dwellinghouse than one of the walls which not only contains and encloses it but is part of its very fabric. The wall stands partly on land immediately next to other land which undoubtedly forms part of the curtilage and in truth it forms part and parcel of the building. The dwellinghouse simply could not function as such without it, whatever the ownership position. Remove it and in all probability, the remainder of the back addition would collapse.

(9) A conclusion that the wall falls within the curtilage of No. [--] would almost inevitably mean that the two adjoining curtilages overlap each other. I can see no rea-son why that should not be so, however, as similar considerations would arise and because such a small area would be involved, extending only to the far side of the wall in each instance. That would not by itself imply any rights of access or to carry out works to the wall without the adjoining owner's consent. The existence of legisla-tion such as The Party Wall Act is an acknowledgement of the needs and rights of adjoining occupiers as well as a means of securing and protecting them. It is thus a further indication and indeed statutory recognition of the 'intimate association' referred to above. Moreover, although very different facts were involved, it was held in the *Sutcliffe* case, above, that one building and its curtilage may fall within the cur-tilage of another building. Taking account of the above authorities and the particular circumstances in this case, ownership in my view is but one factor to be considered, and does not determine the matter exclusively.

Conclusion

(10) In terms of its functional relationship to the building and the area it occupies, I conclude that the party wall falls within the curtilage of the dwellinghouse for the

purposes of the GPDO, even if that might not be so, for example, in conveyancing terms. It follows that to extend it in the manner specified would fall within the GPDO provisions. I have considered all other matters raised, but in the absence of any other objection, none serves to overturn this conclusion and the appeal should thus be allowed.

Formal decision

(11) In exercise of the powers transferred to me, I allow the appeal and I attach to this decision a Certificate of Lawfulness describing the proposed operation which I consider lawful.

(signed)

R O Evans

Planning Inspectorate Appeal Decision dated 11 October 2001 by R O Evans BA (Hons) Solicitor, an inspector appointed by the Secretary of State for the Environment, Transport and the Regions

Glossary

AAC Autoclaved aerated concrete (block).

AD Approved Document (England and Wales).

ALS/CLS Surfaced timber with rounded arrises widely used for studwork in timber frame construction. Common sizes 38 × 89 mm and 38 × 140 mm. (American Lumber Standards Board of Review, Canadian Lumber Standards Board of Review).

BRE Building Research Establishment Ltd.

BUR Built-up roofing. Widely used flat-roof weathering system, with membranes manufactured using polyester reinforcement and modified bitumen.

Back addition Rearward projection in 19th and early 20th century terraced dwellings. The back addition may have the same number of storeys as the main building but is subordinate in width. A back addition is generally shared between a pair of houses and is thus divided along its length by a party wall; the roof often comprises a pair of independent lean-to slopes (see figure below). The back addition roof space is sometimes incorporated as part of a larger conversion.

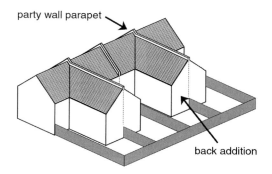

Back addition

Bedded verge Roof edge finishing detail for gable ends. An oversailing rigid undercloak is formed before battens and roofing material are fixed. The void between the undercloak and the roofing material is then filled with mortar and the bedding struck off cleanly.

Breather membrane Allows water vapour to escape from the building envelope but prevents the passage of liquid water. Approved Document C indicates that breather membrane, rather than traditional impermeable underlay, should be used behind vertical tile hanging.

Certificate of Lawfulness (CoL) Generally used to establish the lawfulness of building work carried out under permitted development legislation. Sometimes called a Lawful Development Certificate (LDC).

Cloaked verge   Verge formed at gable ends with purpose-made L-shaped tiles or slates with a leg returned over the head of the wall. Does not require mortar.

Coffin tank   The coffin tank, or loft tank, is widely used in conversions where a conventional water tank cannot be accommodated. As its name suggests, it is a long and relatively shallow tank that is suitable for fixing in confined spaces such as the eaves or the apex of the roof.

Coffin tank (Image courtesy of Polytank Group Ltd)

Coffin tank typical volumes and dimensions

	114 litres (25 gallons)	227 litres (50 gallons)	318 litres (70 gallons)	454 litres (100 gallons)
Length (mm)	1390	1640	1650	1670
Width (mm)	500	450	460	690
Height (mm)	**310**	**485**	**475**	**580**

Dead load   Load due to the weight of walls, permanent partitions, floors, roofs and finishes including services and all other permanent construction.

ELV   extra low voltage

EPBD   Energy Performance of Buildings Directive.

EPDM   Synthetic sheet rubber membrane used as a roofing material.

Existing opaque fabric   Alternative expression for *retained thermal element* (omitted from final version of ADL1B).

FFL   Finished floor level (generally on drawings).

Fenestration   The arrangement of windows in a building.

Firring   Firring pieces are tapered timber elements that are nailed along the tops of roof joists (or at right angles to them in some instances) in order to achieve a fall across a flat roof. Generally these are cut to achieve a fall of about 1:40.

Fixed building services   These include heating systems, hot water systems, fixed internal and external lighting, cooling systems and mechanical ventilation systems.

Floor-to-floor   The measurement between the surfaces of different floors (e.g. an existing upper floor and the new floor in a loft conversion). This measurement is critical when commissioning a staircase for a loft conversion.

Friction fit   A push fit achieved by slight over-sizing of the material to be fitted (e.g. rigid insulation boards between rafters or studs).

GPDO   The Town and Country (General Permitted Development) Order 1995. Planning legislation. Most loft conversions in England and Wales are carried out under permitted development rights set out in the GPDO.

GRP   Glass reinforced plastic (fibreglass). Sometimes used as weathering for small dormers. Often pigmented to resemble lead or zinc.

Gable parapet   Projection of a gable wall above the roof surface to the flank side of a building (see also *party wall parapet*).

Gallows bracket   A welded mild steel bracket formed from rolled steel angles. These are often used as support brackets where an internal chimney breast has been removed (see also Chapter 5). Usually secured to walls with expansion bolts. Gallows brackets are not universally accepted by building control services and are generally not used if the wall supporting the bracket is part of a flue.

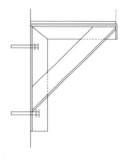

Gallows bracket

Glulam   Glued laminated timber.

HSFG   High-strength friction grip (bolt).

Hip iron   A scroll-ended strap screwed to the base of the hip rafter to provide support for hip tiles. Sometimes called a hip hook. Generally made from 6 mm galvanised steel. Earlier examples are wrought iron.

House Longhorn beetle (*Hylotrupes bajulus*)   Capable of causing severe damage to softwoods, particularly in roof structures in certain districts. The beetle is black or dull brown in colour and between 10 and 20 mm in length. Emerges from infected timber between July and October. Larvae are up to 30 mm in length. Emergence holes are oval in shape and about 6–10 mm.

Guidance in Approved Document A *Structure* indicates that softwood timber for roof construction, including ceiling joists, should be adequately treated to prevent infestation. This would also include, for example, vertical studs in a dormer wall. In general, the guidance is satisfied by the use of tanalised timber. Further guidance on suitable preservative treatments is provided in the British Wood Preserving and Damp-proofing Association's manual.

Areas at risk from House Longhorn beetle
- Borough of Bracknell Forest (parishes of Sandhurst and Crowthorne)
- Borough of Elmbridge
- District of Hart (parishes of Hawley and Yateley)

- District of Runnymede
- Borough of Spelthorne
- Borough of Surrey Heath
- Borough of Rushmoor (area of the former district of Farnborough)
- Borough of Woking

Imposed load The load assumed to be produced by occupancy or use, including the weight of moveable partitions and snow. Excludes wind loads. Also described as live load.

Interstitial condensation Deposition of liquid water, from vapour, if it occurs within or between elements in the building (e.g. water that condenses within insulation).

LDC Lawful Development Certificate (planning). See *Certificate of Lawfulness* above.

Lamp A light source (bulb). A lamp may be incandescent or fluorescent.

Light fitting/lighting fitting (energy-efficient) Both expressions are used in Approved Document L1B. In the context of efficient electric lighting, a lighting fitting includes the lamp, control gear and an appropriate housing, reflector, shade or diffuser or other device for controlling the output of light. An energy-efficient light fitting must only take lamps having a luminous efficacy greater than 40 lumens per circuit-watt. Circuit watts refers to the power consumed in lighting circuits by lamps and their associated control gear and power factor correction equipment.

Live load (see *imposed load*)

LOLER Lifting Operations and Lifting Equipment Regulations (1998). Legal requirements relating to the use of lifting equipment.

Low-E glass Low-emissivity glass allows heat and light to pass into a building but limits heat loss from the building. This is achieved by applying a thin metallic coating to one side of the glass during manufacture. There are two principal types of low-E glass: hard (emissivities in the range 0.15 to 0.2 ε_n) and soft (emissivities in the range 0.05 to 0.1 ε_n). By comparison, the emissivity of plain glass is about 0.89.

Luminaire Generally, a complete lighting assembly including lamp, lampholder and associated housing.

MoE Means of escape (from fire). Usually refers to an emergency egress window.

MS Mild steel (on drawing).

Material alteration (Building Regulations) Material alteration is defined in regulation 3 of the Building Regulations 2000 (as amended). An alteration is material, for example, if it would result in a building or controlled service not complying with a relevant requirement where previously it did. A material alteration would also include making a sub-standard part of an existing building worse than it already was. Note that the scope of material alteration changes from time to time and, at the time of writing, its definition is spread across a number of amendments to the Building Regulations.

Material change of use Regulation 5 of the Building Regulations 2000 (as amended). There is a material change of use where there is a change in the purposes for which, or the circumstances in which, a building is used. Note that a loft conversion in single-family dwelling is not a material change of use. Material changes of use would include the conversion of a building into a dwelling (where previously it was not) and the conversion of a house into flats (where, for example, it was previously a single-family dwellinghouse).

ODPM Office of the Deputy Prime Minister. The government department responsible for Building Regulations and Approved Documents in England and Wales.

OPC Ordinary Portland cement.

OSB Oriented strand board. In loft conversions, principally used for external sheathing and roof decking.

PD Permitted development (see Chapter 1).

PFC Parallel flange channel (structural steel).

PIR Polyisocyanurate (rigid thermal insulation boards).

PUR Polyurethane (rigid thermal insulation boards).

PWA/PWeA Party Wall etc. Act 1996.

Party wall parapet Projection of the party wall above the plane of the roof between terraced or semi-detached dwellings, usually a minimum of 400 mm. Party wall parapets were once a regulatory requirement and were intended to inhibit the spread of fire between adjacent buildings. Widely used throughout the 19th century. See also *gable parapet*.

Plain tile Plain tiles are widely used as vertical cladding in loft conversions. Traditionally, plain tiles were made from clay, but concrete is now widely used in their manufacture. The plain tile represents perhaps the earliest example of standardisation in English construction, its size set at $10^1/_2$" $\times 6^1/_4$" by an enactment of 1477. Under the current standard, plain tiles are 265×165 mm.

Proof load The proof load of a bolt is the specified load it must withstand without permanent set.

RMI Repair, maintenance and improvement (of buildings).

RSJ Rolled steel joist. In strict terms, this is a designation in its own right (although seldom used now) and is distinct from the universal beam or column. However, it is widely used as a generic term to describe any steel I-beam.

RWP Rainwater pipe (on drawings).

Renovation (of a thermal element) For the purposes of Part L, this is defined as 'the provision of a new layer in the thermal element or the replacement of an existing layer, but excludes decorative finishes'. Renovation of a thermal element acts as a trigger for upgrading. Paragraph 54 of ADL1B indicates that, where more than 25% of the surface area of a thermal element is being renovated, a way of showing compliance with the regulation would be to upgrade the whole of the element in question. Note that re-tiling and re-felting an existing roof slope would constitute a renovation.

Replacement thermal element The expression *replacement thermal element* is used in ADL1B. A replacement thermal element could be defined as a wall, floor or roof that is removed in its entirety during building work and replaced with new construction at the original position in the building envelope. In the case of a loft conversion, a front roof slope that is completely dismantled and replaced with new rafters, battens, underlay and covering would constitute a replacement element. However, re-tiling and re-felting an existing roof slope would constitute a renovation (see below).

Retained thermal element   This expression is used in Approved Document L1B. A retained thermal element may be defined as a wall, floor or roof that becomes part of the building's thermal envelope where it was not before. Existing roof slopes and gable walls incorporated in a loft conversion would normally qualify as retained thermal elements.

Roof light/window   A window set in the same plane as the roof.

SAP   Standard Assessment Procedure.

SBSA   Scottish Building Standards Agency.

SEDBUK   Seasonal Efficiency of Domestic Boilers in the UK.

SPG   Supplementary planning guidance.

SVP   Soil and vent pipe (on drawings).

Simple payback   For the purposes of Approved Document L1B, this is the marginal additional cost of implementing an energy efficiency measure (excluding VAT) divided by the value of the annual energy savings achieved by that measure. For example, if the additional cost of implementing a measure (in materials and labour) was £430 and the value of the annual energy savings was £38 per year, the simple payback would be: $430 \div 38 = 11.3$ years.

Skillings   The name given to a sloping ceiling formed by a roof, for example, where the rafters of a pitched roof are boarded over to form a ceiling in a loft conversion.

Swept valley   A curved valley formed in tile, stone or slate rather than metal flashing.

Thermal element   Thermal element means a wall, floor or roof which separates the internal conditioned space from the external environment, and includes all parts of the element between the surface bounding the internal conditioned space and the external environment. It would be reasonable to assume that this would include the walls between the heated part of a loft conversion and an unheated roof void in an adjoining building.

TRADA   Timber Research and Development Association.

Triangulation   The creation of stable triangular configurations, particularly in roof structures. Distinct from triangulation in surveying/cartography.

UDP   Unitary Development Plan

U-**value**   The amount of heat energy (in watts) transmitted through one square metre of building (e.g. a wall) for every one-degree difference (kelvin) between external and internal temperature – W/m^2K.

VCL   Vapour control layer (see below).

Vapour check   A vapour impermeable membrane, often heavy-gauge polythene sheet with taped joints, used to prevent moisture-laden air from within a building escaping into and condensing within its structural fabric (see also *interstitial condensation*). Sometimes referred to as a VCL or vapour barrier). Some plasterboards and rigid-sheet insulation material have an integral vapour check.

WBP   Plywood – weather and boil proof. Widely used in sheathing and decking. Sometimes described as WPB.

Bibliography and useful contacts

Primary references and standards

BS 449-2:1969 *Specification for the use of structural steel in buildings*
Note: this standard, while current, is no longer referenced in Approved Document A, *Structure.*
BS 5250:2002 *Code of practice for control of condensation in buildings*
BS 5268-2:2002 *Structural use of timber. Code of practice for permissible stress design, materials and workmanship*
BS 5588-1:1990 *Fire precautions in the design, construction and use of buildings. Code of practice for residential buildings*
BS 5628-1:1992 *Code of practice for use of masonry. Structural use of unreinforced masonry*
BS 5839-6:2004 *Fire detection and fire alarm systems for buildings. Code of practice for the design, installation and maintenance of fire detection and fire alarm systems in dwellings*
BS 5950-1:2000 *Structural use of steelwork in building. Code of practice for design. Rolled and welded sections*
BS 9251:2005 *Sprinkler systems for residential and domestic occupancies. Code of practice*
BS EN ISO 13788:2002 *Hygrothermal performance of building components and building elements*
Eurocode 3: *Design of steel structures*
Eurocode 5: *Design of timber structures*
TRADA Technology Ltd (2004) *Span Tables for Solid Timber Members in Floors, Ceilings and Roofs (excluding trussed rafter roofs) for Dwellings*
BRE (2005) *The Government's Standard Assessment Procedure for Energy Rating of Dwellings (SAP 2005)* Published on behalf of DEFRA.
NBS (2006) *Domestic heating compliance guide.* Published on behalf of the ODPM.
EST (2006) Low energy domestic lighting. GIL20.

Publications and technical literature

Clay Roof Tile Council (2004) *A Guide to Plain Tiling Including Vertical Tiling.*
Yeomans, D. (1997) *Construction Since 1900: Material.* Batsford.
BRE (2006) *Conventions for U-value Calculations.* BRE Report 443.
Brunskill, R. & Clifton-Taylor, A. (1977) *English Brickwork.* Ward Lock Ltd.
TRADA Technology Ltd (2005) *Fire Resisting Doorsets by Upgrading.* WIS 1–32.
BRE (1988) *Increasing the Fire Resistance of Existing Timber Floors.* BRE Digest 208.
BRE (1998) *Timbers: Their Natural Durability and Resistance to Preservative Treatment.* BRE Digest 429.
Goring, L. (1998) *Manual of First- and Second-Fixing Carpentry.* Elsevier.
Brick Development Association (2001) *Observations on the Use of Reclaimed Clay Bricks.* PBM 1.4.
Lead Sheet Association (2003) *Rolled Lead Sheet – The Complete Manual.*
Mindham, C.N. (1999) *Roof Construction and Loft Conversion.* Blackwell.
Cobb, F. (2004) *Structural Engineer's Pocket Book.* Elsevier.
Brick Development Association (2000) *The BDA Guide to Successful Brickwork.* Butterworth-Heinemann.

Billington, M.J., Simons, M.W. and Waters, J.R. (2004) *The Building Regulations Explained and Illustrated*. Blackwell Publishing.
ODPM (2002) *The Party Wall etc. Act 1996: Explanatory Booklet*.
BRE (2002) *Thermal Insulation: Avoiding Risks*. BRE Report 262.
Brunskill, R.W. (1985) *Timber Building in Britain*. Victor Gollancz Ltd.
TRADA Technology Ltd (2001) *Timber Frame Construction*.

Regulatory policy

Building Regulations: England and Wales
Office of the Deputy Prime Minister (ODPM)
Buildings Division, Eland House
Bressenden Place
London SW1E 5DU
Tel: 020 7944 4400
www.odpm.gov.uk

Building Regulations: Northern Ireland
Department of Finance and Personnel
Building Regulations Unit, OBD
River House
48 High Street
Belfast BT2 8AA
Tel: 028 9025 7326
www2.dfpni.gov.uk/buildingregulations

Building Regulations: Scotland
Scottish Building Standards Agency (SBSA)
Denholm House
Almondvale Business Park
Livingston EH54 6GA
Tel: 01506 600400
www.sbsa.gov.uk

Contacts

Association of Building Engineers (ABE)
Lutyens House
Billing Brook Road
Weston Favell
Northampton NN3 8NW
Tel: 01604 404121
www.abe.org.uk

Association for the Conservation of Energy (ACE)
Westgate House
Prebend Street
London N1 8PT
Tel: 020 7359 8000
www.ukace.org

Brick Development Association Ltd (BDA)
Woodwide House
Winkfield
Windsor
Berkshire SL4 2DX
Tel: 01344 885651
www.brick.org.uk

British Standards Institution (BSI)
389 Chiswick High Road
London W4 4AL
Tel: 020 8996 9000
www.bsi-global.com

British Wood Preserving & Damp-proofing Association (BWPDA)
1 Gleneagles House
Vernongate
Derby DE1 1UP
Tel: 01332 225100
www.bwpda.co.uk

Building Centre
26 Store Street
London WC1E 7BT
Tel: 020 7692 4000
www.buildingcentre.co.uk

Building Research Establishment (BRE)
Bucknalls Lane
Garston
Watford WD25 9XX
Tel: 01923 664000
www.bre.co.uk

Clay Roof Tile Council
Federation House
Station Road
Stoke-on-Trent ST4 2SA
Tel: 01782 744631
www.clayroof.co.uk

Construction Fixings Association
I.S.T. Henry Street
Sheffield S3 7EQ
Tel: 0114 278 9143
www.fixingscfa.co.uk

Copper Development Association (CDA)
5 Grovelands Business Centre
Boundary Way
Hemel Hempstead
Hertfordshire HP2 7TE
Tel: 01442 275700
www.cda.org.uk

Corus Construction Centre
PO Box 1
Brigg Road
Scunthorpe
DN16 1BP
Tel: 01724 404040
www.corusconstruction.com

Energy Saving Trust (EST)
21 Dartmouth Street
London SW1H 9BP
Tel: 020 7222 0101
www.est.org.uk

Federation of Master Builders
Gordon Fisher House
14–15 Great James Street,
London WC1N 3DP
Tel: 020 7242 7583
www.fmb.org.uk

Fire Protection Association (FPA)
London Road
Moreton-in-Marsh
Gloucestershire GL56 0RH
Tel: 01608 812500
www.thefpa.co.uk

Glued Laminated Timber Association (GLTA)
Chiltern House
Stocking Lane
High Wycombe HP14 4ND
Tel: 01494 565180
www.glulam.co.uk

Institution of Structural Engineers
11 Upper Belgrave Street
London SW1X 8BH
Tel: 020 7235 4535
www.istructe.org.uk

LABC Services
137 Lupus Street
London SW1V 3HE
Tel: 020 7641 8737
www.labc-services.co.uk

Lead Sheet Association (LSA)
Hawkwell Business Centre
Maidstone Road
Pembury
Tunbridge Wells TN2 4AH
Tel: 01892 822773
www.leadsheetassociation.org.uk

London District Surveyors Association (LDSA)
www.londonbuildingcontrol.org.uk

National Building Specification Ltd (NBS)
The Old Post Office
St. Nicholas Street
Newcastle upon Tyne NE1 1RH
Tel: 0845 456 9594
www.thenbs.com

National House-Building Council (NHBC)
Buildmark House
Chiltern Avenue
Amersham
Buckinghamshire HP6 5AP
Tel: 01494 735363
www.nhbc.co.uk

Pyramus & Thisbe Club
Rathdale House
30 Back Road
Rathfriland BT34 5QF
Tel: 028 4063 2082
www.partywalls.org.uk

Royal Institute of British Architects (RIBA)
66 Portland Place
London W1B 1AD
Tel: 020 7580 5533
www.riba.org

Royal Institution of Chartered Surveyors (RICS)
12 Great George Street
London SW1P 3AD
Tel: 0870 333 1600
www.rics.org

Steel Construction Institute (SCI)
Silwood Park
Ascot
Berkshire SL5 7QN
Tel: 01344 623345
www.steel-sci.org

Timber Research and Development Association (TRADA)
Stocking Lane
Hughenden Valley
High Wycombe
Buckinghamshire HP14 4ND
Tel: 01494 569600
www.trada.co.uk

Trussed Rafter Association (TRA)
31 Station Road
Sutton cum Lound
Retford
Nottinghamshire DN22 8PZ
Tel: 01777 869281
www.tra.org.uk

Index